"Andersen and Evia provide a comprehensive, yet accessible, explanation of the various categories of the content discipline. More importantly, they outline what it will take to prepare students for the kind of work going on right now in the workplace. Their approach is to teach carefully and specifically. This book will interest both academia and industry."

Tracy Bridgeford, *University of Nebraska at Omaha, United States*

"*Technical Communication and the Discipline of Content* is exactly what the field of technical communication needs right now—rigorous empirical research that describes both how content work is evolving and what those shifts mean for educators, students, and practitioners of content strategy."

Erin Friess, *University of North Texas, United States*

"Andersen and Evia share novel research into, and apply thoughtful reflection on, the core question of how the education of technical communicators needs to evolve especially now that AI has joined us at the table. For those of us working in this field, the time to engage this question is now. This book will help us to do this."

Joe Gollner, *Managing Director of Gnostyx Research Inc., Canada*

TECHNICAL COMMUNICATION AND THE DISCIPLINE OF CONTENT

This book explores how workflows and technologies that treat content as computable data are changing the roles, work activities, and outputs of professional technical communicators.

It describes how the need for disciplinary approaches to design, manage, and deliver content has given rise to "the discipline of content"—content strategy, content design, content engineering, content operations—that increasingly defines a facet of technical communication work in modern organizations. This book draws on extensive research of the discipline of content and dozens of interviews with industry leaders, hiring managers and academic administrators, educators, and alumni. These first-hand accounts outline how roles and activities in content organizations are changing, how these changes are impacting hiring needs and practices, and what skills and qualities students and early-career professionals now need to obtain content-related jobs and advance to strategic positions. This book also offers guidance for building curricular pathways that prepare students for work in the discipline of content and offers strategies for enhancing pathways through industry outreach and partnerships.

A thorough assessment of the implications of the discipline of content for technical communication, this book will be of interest to students and scholars in the fields of technical writing, professional and public writing, content strategy, content marketing, and information design.

Rebekka Andersen is an Associate Professor and the Associate Director for Professional Writing in the University Writing Program at the University of California, Davis, United States.

Carlos Evia is a Professor, Associate Dean, and Chief Technology Officer in the College of Liberal Arts and Human Sciences at Virginia Tech, United States. He is the author of *Creating Intelligent Content with Lightweight DITA* (Routledge, 2018).

ATTW Series in Technical and Professional Communication

Michele Simmons and Lehua Ledbetter, Series Editors

Augmentation Technologies and Artificial Intelligence in Technical Communication: Designing Ethical Futures
Ann Hill Duin and Isabel Pedersen

Sustainable Crowdfunding: Research-Based Analysis for Communication Campaigns
Adam R. Pope

Designing for Social Justice: Community-Engaged Approaches in Technical and Professional Communication
Jialei Jiang and Jason C. K. Tham

Technical Communication and the Discipline of Content: Considerations for Research, Training, and Career Readiness
Rebekka Andersen and Carlos Evia

For additional information on this series please visit www.routledge.com/ATTW-Series-in-Technical-and-Professional-Communication/book-series/ATTW, and for information on other Routledge titles visit www.routledge.com.

TECHNICAL COMMUNICATION AND THE DISCIPLINE OF CONTENT

Considerations for Research, Training, and Career Readiness

Rebekka Andersen and Carlos Evia

Routledge
Taylor & Francis Group
NEW YORK AND LONDON

Designed cover image: Red Deer/Shutterstock

First published 2026
by Routledge
605 Third Avenue, New York, NY 10158

and by Routledge
4 Park Square, Milton Park, Abingdon, Oxon, OX14 4RN

Routledge is an imprint of the Taylor & Francis Group, an informa business

ISBN: 9781032588476 (hbk)
ISBN: 9781032588469 (pbk)
ISBN: 9781003451730 (ebk)

DOI: 10.4324/9781003451730

Typeset in Sabon
by codeMantra

CONTENTS

FIGURES

TABLES

BOXES

PREFACE

Content as Key and Catalyst

This book documents the many ways in which workflows and technologies that treat content as computable data are changing the roles, activities, and outputs of technical communicators working in organizations that have undergone digital content transformations. It captures the results of our research with the main purpose of describing how the need for standardized, measured approaches to design, manage, and deliver content has given rise to *the discipline of content*, comprising four disciplinary areas: content strategy, content design, content engineering, and content operations. Through extensive research of the discipline of content and dozens of interviews with leaders in the content industry, administrators and educators in academia, and alumni of technical communication programs, as well as a survey of hiring managers, in this book, we sort out definitions and disciplinary boundaries of content and explore what these changes mean for hiring needs and for training and education in the field of technical communication.

Anyone who has a stake in the field and an interest in ways in which technical communication and the discipline of content intersect has much to gain from reading this book. Whether you are an entry- or mid-level practitioner, manager, recruiter, professional looking for a career change, trainer, administrator, educator, graduate student, or undergraduate student, this book offers you an in-depth understanding of how roles, activities, and hiring needs are changing in organizations that treat content as a valued asset and computable data. If you are an administrator or educator in a technical communication academic program, you will also find, in addition to a competencies framework for content professionals, guidance on how you might innovate at the program and course level to prepare students for content jobs and career mobility while also keeping an eye toward program sustainability. If you are a practitioner, manager of a technical content team, or industry leader, you will find strategies for building partnerships with academic programs and participating in initiatives and activities that support students' development of competencies and engagement in experiential learning. You will further find ideas for better defining the entry-level content-focused position as well as career and advancement paths within the content discipline.

Proficiency working with technical content is clearly becoming a key competency for technical communication professionals—a non-negotiable for entry-level and more advanced content-focused roles. As such, it's a catalyst for change in technical communication—both in academic programs and curricula, and in organizational onboarding and training programs. At the core of our field's collective identity—the through-line connecting all of its different facets—is human-centered design. Stakeholders in the field have much to gain by engaging with modern operations that optimize content production and delivery for the diverse, complex, and demanding human users who need effective technical communication products. And administrators and educators have much to gain by creating curricular spaces that prepare students for careers as strategists, architects, and other important roles in organizations that work in the discipline of content.

ACKNOWLEDGMENTS

This book is the main deliverable of a research project that started as a conversation on discrepancies between what a sector of the technical communication industry perceives as its core mission and what academic programs in this discipline are actually teaching at the undergraduate level. The conversation evolved throughout the years into a structured research project that has taken us to many conference presentations, magazine articles, podcast interviews, and finally this book. And we do mean years: the COVID-19 pandemic, family hardships, failed (and successful) grant applications for research support, and new job responsibilities and challenges surely made this project a bit longer than it should have been. However, the long process also allowed us to become better collaborators and friends.

Many colleagues in industry and academia have supported our work and helped us along the way. The list is long and will surely fail to acknowledge all our supporters, cheerleaders, and even critics. We want to recognize the influence and help of Scott Abel, Rahel Bailie, Tatiana Batova, Ryan Boettger, Patrick Bosek, Tracy Bridgeford, Stan Doherty, Erin Friess, Joe Gollner, Bill Hart-Davidson, Alan Houser, Jimmy Ivory, Larry Kunz, Chris Lam, Tim Lockridge, Sarah O'Keefe, Laura Palmer, Kelly Pender, Dawn Stevens, Jason Swarts, Jason Tham, Russell Willerton, and our anonymous research participants from industry and academia. We want to also thank the Center for Information-Development Management for its support of this project, from inviting us to present at conferences at different stages of the project to helping us distribute our survey of technical communication managers far and wide. Roxanne Elliott, PhD student in Education at UC Davis, helped us analyze our interviews of undergraduate program alumni; we thank Roxanne for her astute insights and careful analysis. We also thank the UC Davis academic senate for the several small grants that supported Roxanne and other aspects of our project. We are grateful, too, for the support of Michele Simmons and Lehua Ledbetter; as editors of the *ATTW Series in Technical and Professional Communication*, they saw the potential of this project when we were considering writing a book and offered us helpful feedback on drafts throughout our planning and writing process.

Rebekka wants to thank her family for their patience and support and especially Ken for his encouragement, which sometimes would come in the form of reminders of all the adventures they would go on "when the book is done." She thanks Carlos for his wisdom, expertise, and optimism, without which this book would not have been possible. She is grateful for the opportunity to have worked with such a kind and thoughtful collaborator. Her colleagues Kory Lawson Ching, Dana Ferris, Dan Melzer, Trish Serviss, and Carl Whithaus also offered helpful feedback and perspective on early drafts of the book proposal and sample chapters. She thanks them for their support.

Carlos wants to thank Jane and Sofía for their love and support. Particularly, he thanks Rebekka for being the tireless engine behind this book and the long research project behind it. As the unquestionable leading author of this book, Rebekka organized, analyzed, wrote, and edited over the years, supporting and dealing with Carlos through good and bad times in our professional lives.

We dedicate this book to JoAnn Hackos and Carolyn Rude for their guidance and mentorship.

SECTION I

Discrepancies, Challenges, and Opportunities

INTRODUCTION

The Need for Technical Communicators as Content Professionals

In the fall of 2017, we participated in a panel discussion at a conference organized by the Center for Information-Development Management (CIDM), a professional organization that brings together managers of information development teams to facilitate the sharing of information about current trends and best practices. The panel, titled "The State of Education in Technical Communication," had us as representatives of the academic side of the field in a conversation with colleagues from the industry side of technical communication.

That afternoon in Vermont, as we provided an overview of the many academic courses, programs, and degrees available in the United States with titles related to technical communication, our counterparts from industry seemed a little surprised. Some of them were in managerial positions and were frequently looking for new hires. They claimed that they could not find qualified new workers and, as a result, ended up hiring the same "dinosaurs" from the field, who moved from one company to the other. Therefore, our numbers and figures about the available pool of graduates with degrees in technical communication were not matching with the perspective of those who wanted to hire recent graduates and could not find them. A colleague attending that panel added that her company used to hire fresh college graduates, but that the onboarding process had become too long and costly to maintain; technical communicators with new degrees were often unfamiliar with new approaches to developing

DOI: 10.4324/9781003451730-2

technical content and had to catch up with many tools, processes, and workflows at that company. Our industry colleagues at that conference session had one thing in common: they needed technical communicators as *technical content professionals*—that is, individuals comfortable working with what Rebekka has described as "structured content that is highly adaptable and portable and can be configured on the fly in response to specific user requests" (Andersen, 2014, p. 116).

Those mismatched expectations involving technical communication and representations of content were not new to us. As researchers and educators in the field of technical communication, we have long been interested in the ways in which technical communication is defined and practiced in both academia and industry. It has been a priority for us to explore research questions that grow out of practice and to bring our knowledge of practice into our classrooms and publications. The adoption of workflows and technologies that treat (mainly written) content as computable data, and more recently data that could be simultaneously enhanced and hampered by generative artificial intelligence tools, arguably represents the greatest shift in the work activities and outputs of professional technical communicators over the past 20 years. Business requirements and user needs and contexts have changed considerably, and now content in a growing number of organizations is being designed and orchestrated for a multitude of devices, tools, and delivery channels. This means that many technical communicators today contribute to content ecosystems that rely on modular components—the essential building blocks for the vast array of information products that organizations deliver in a variety of publishing formats, such as printed documentation, websites, embedded user assistance, dynamic delivery, and mobile applications.

To stay abreast of these changes and the pressing questions and challenges facing technical content teams, we have each strived to attend one or two industry conferences a year (e.g., CIDM Best Practices, LavaCon, the Society for Technical Communication [STC] Summit); participate on industry advisory committees and boards (e.g., the DITA Technical Committee, the CIDM Advisory Council); follow online content-focused discussions happening in social media, blogs, and webinars; and establish connections that eventually became collaborations with well-regarded industry leaders in the technical content workplace, such as Rahel Bailie, JoAnn Hackos, and Sarah O'Keefe. As presenters and active participants at industry conferences, we have also engaged in many conversations with technical communication professionals working as managers, writers, strategists, architects, analysts, consultants, and other roles. These conversations have often centered on research needs and questions about how academic programs in technical communication are preparing students for careers in the content-focused sector of the field.

Our shared interest in implications of contemporary content practices for research and education in the field brought us together in 2016, first when Carlos invited Rebekka to give a talk at Virginia Tech, and second when Rebekka emailed Carlos from the CIDM Best Practices Conference with the subject line: "Feeling Energized!" Her initial email expressing excitement and optimism for potential research collaborations with industry colleagues (senior managers and their information teams) became a series of threads and conversations over the next year about discrepancies that we were noticing between conversations happening in industry and conversations happening in academia. One of those threads went like this:

> I felt like I was in a science fiction movie while attending sessions at Best Practices. Joe Gollner was talking about Content 4.0—what's coming. As he was talking, I kept thinking about how most of our programs are still teaching Content 1.0. A handful are teaching Content 2.0. And almost none have embraced Content 3.0, which is where the TC industry is now. The good news, I think, is that TC industry managers are very interested in collaborating with us to help improve our programs (and to building a body of research that will also benefit them). I'm not sure how we collaborate on rethinking/revamping TC curricula in the field, but I do know that it's going to take a bunch of minds coming together and taking action.

This thread points to two key discrepancies that are at the core of this book's *raison d'être*.

First, in the industry side of technical communication, the topic of "content" and all its evolving flavors and associated practices was dominating conversations happening at conferences, on social media and the web, and in practitioner-focused publications (e.g., books, newsletters, whitepapers). Consultants, managers, and thought leaders were all talking about changing business requirements for content, maturing content technologies, and strategic approaches for designing content experiences. The concept of **Content 4.0** was also gaining attention—the idea being that content innovations must evolve along with industrial innovations. According to its proponents, content had evolved from the book or document (when content and format were intertwined) (**Content 1.0**) to the single-sourceable modular topic (when content and format were separated) (**Content 2.0**) to the modular component (when component content from multiple departments was integrated) (**Content 3.0**). Looking forward, thought leaders like Joe Gollner (2016) were strategizing on how to engineer smart content objects and molecules that could assemble, transform, and render "dynamically in response to contexts and agent needs" (to learn more about **Content 4.0**, refer also to Technically Write IT (2018) and Simple[A] (2019).

Despite these conversations being front and center in industry, however, "content" was not a topic receiving much attention in technical communication academic circles in the United States at that time. Content was neither a central topic in our curricula (see, e.g., Melonçon & Henschel, 2013) nor at our conferences. Although a few special journal issues had been dedicated to exploring the move to modular content in industry (a 2008 special issue of *Technical Communication Quarterly*, and a 2015 and 2016 special issue of *IEEE Transactions on Professional Communication*), published academic research on content practices was relatively scarce (and it remains scarce today, with a handful of special issues and edited collections having been published since 2018).

Second, as we mentioned in the opening of this introduction, in our interactions with managers whom we had met at conferences and worked with on committees, we were hearing a consistent message that young people in the field were needed but difficult to find. One manager had even commented, "I cannot find any qualified young people. Where are all the young people?" The assumption seemed to be that college graduates specialized in other things and were not interested in this area of technical communication. We were surprised by how few managers were aware that full-fledged undergraduate degree programs in technical (and professional) communication even existed in the United States. And many of those who did recruit from undergraduate programs tended to comment on the challenge of hiring new graduates who could adapt quickly and run with a project. Yet, at the same time that we were hearing these concerns, we knew that undergraduate degree programs in technical and professional communication had increased by 131% between 2005 and 2012 (see Melonçon & Henschel, 2013). And by many accounts, these graduates were getting jobs. Why were content-oriented managers struggling to find qualified young technical communicators when we were graduating more aspiring technical communication professionals than ever?

Around the time when we were noticing these discrepancies, Miles Kimball (2016) published "The Golden Age of Technical Communication." In that article, he described the current era of technical communication as the Glass Age, characterized by the separation of form and content and the separation of strategic decision-making from writing. These separations, he suggested, meant that writers in organizations that have adopted workflows and technologies that treat content as format-free modules were no longer positioned as the rhetorical agents they once were:

> But the separation of form and content has unfortunately also separated many professional writers from this traditional role. They no longer design an integrated document in which they strategically apply visual

> and lexical rhetoric to solve human problems. Instead, the work of a professional technical writer has increasingly devolved into a more mundane task: Pounding out fragmented chunks of prose that will be assembled by a system designed by someone else. This movement toward separating content writing from strategic design makes technical writing of this type look less like a profession and more like a modular, relatively mechanical activity—just like any part of a machine. *(p. 9)*

Kimball argued that only those writers who were able to move into strategic design roles would be positioned to exert meaningful agency in their organizations. But jobs for writers in these roles, he suggested, were not nearly as abundant as the less-well-paid "mundane" jobs subject to automation. In Kimball's view, the job outlook for technical writing professionals as a whole was bleak, leading him to conclude that undergraduate programs should move away from preparing students for work in the profession, or what he called "strategic" technical communication.

Others had painted a similarly bleak picture of the outlook for technical writers (refer to, e.g., Virtaluoto, Sannino, & Engestrom, 2016; Swarts, 2010). Writing from their perspective as consultants in industry, Jack Molisani and Scott Abel (2012) suggested that the commodification of technical writing had led to reduced employment opportunities for graduates with degrees in writing, as almost anyone with a college degree could write quality content following predefined rules and standards (p. 15). Rebekka, too, had described how the "schemas and standards embedded in the tools have made it possible for people who are not trained technical writers to produce well-structured content, allowing for the outsourcing and offshoring of writing and editing activities"; the effect for US workers looking for technical writing jobs, she suggested, "is fewer full-time, in-house positions" (Andersen, 2014, p. 121).

Although the commodification of technical writing had certainly increased the number of writing jobs being outsourced, offshored, or automated, we knew from our conversations with managers and consultants working in the content-focused sector of the field that new kinds of writers—that is, technical content professionals—were needed but were difficult to find. The strong employment gains for technical writers predicted by the 2017 editions of the STC Salary Database and the *Occupational Outlook Handbook* seemed to suggest this as well.

In "Keeping Technical Writing Jobs Relevant with Content Automation Technologies," Alex Masycheff (2018), CEO of Intuillion and content technology expert, described these new roles for writers. He suggested that the increased automation of "low-value tasks" meant that technical writers must now "deal with more complicated and creative tasks" and see their job as interdisciplinary, including writing "but also information design, UX, customer support, user engagement, and other areas humans are traditionally good at." Craig Baehr (2015), too, recognized the need for technical communicators to be prepared to take on more active roles. Drawing on the results of a survey that examined manager perspectives on training and education in the field, Baehr reported that these roles meant "increased autonomy, greater management responsibility for information products, new roles, and working on multiple projects simultaneously" (pp. 109–110). For these reasons and others, many undergraduate programs in technical communication now include courses focused on information design and UX.

Given the discrepancies we were seeing between conversations happening in industry and those happening in academic circles, and given what appeared to be a promising job outlook for technical content professionals, we wanted to know how technical communication undergraduate programs could better prepare students for roles that afforded them opportunities for meaningful contributions to content activities in their future organizations. We wanted to know how requirements for entry-level positions in the field had changed such that the managers we had talked with were not able to find qualified young people. And we wanted to know how writers in traditional roles could make the leap to more strategic roles as content professionals. At that time, no published research had specifically examined the entry-level technical content professional position, and we saw that research gap as a real opportunity to contribute new, much-needed knowledge to the field.

This desire to better understand the underexplored and not-well-understood activities and roles of technical communicators in organizations that have adopted workflows and technologies that treat content as computable data seeded a multi-year, multi-staged research project that is the essence of this book. Before describing the project, we turn briefly to sorting out some definitions and disciplinary boundaries of content and explaining why the *discipline of content* is a useful framework for examining changing roles, activities, and hiring needs on technical content teams and their implications for technical communication education and research.

The (Multi)Disciplinary Nature of Content

So far, we have been talking about content as an amorphous concept. The truth is that content is hard to define, and it would be helpful to set some working definitions before returning to our questions introduced earlier in this chapter and our goals for writing this book. Probably, one of the broadest definitions of the term comes from Mike Atherton and Carrie

Hane (2018), who see content as "substantive information that is expressed through a medium" (p. 33). Other long-standing definitions focus on the needs of consumers of digital communication products. Kristina Halvorson and Melissa Rach (2012), for example, provide the following definition in their landmark book *Content Strategy for the Web*:

> Content is what the user came to read, learn, see, or experience. From a business perspective, the content is the critical information the website, application, intranet, or any other delivery vehicle was created to contain or communicate.
>
> *(p. 28)*

Rebekka and our colleague Tatiana Batova (Andersen & Batova, 2015) have explored the definition of content in several publications. Their findings show that content "is used widely to refer to the product- and service-related information that organizations produce for their customers and other stakeholders" (p. 253). Some of the definitions that they quote include the following from Scott Abel (2014):

> "Any text, image, video, decoration, or user-consumable elements" that help people understand "an organization's products or services, stories, and brand."
>
> *(p. 12)*

Abel expands that definition by pointing out that content can be described in the following ways:

- Contextualized data
- The stuff inside a container
- An extension of the user experience
- A business asset (p. 13).

As content becomes more difficult to define in one simple sentence, establishing disciplinary boundaries for professionals working with content in industry and academia turns into a more complicated task. Some colleagues tend to group all content-related activities under the umbrella of content management (mainly when talking about content presented on websites), whereas others have adopted content strategy as a universal key for all content doors. The truth is that content—from "stuff inside a container" to step 1 in an online recipe for *arroz con pollo* to the latest dancing challenge on TikTok—is a complex construct that nowadays should be studied and practiced from several disciplinary lenses (or a combination of those).

Sara Feldman (2020), a technical communicator and content strategist, tried to identify all the disciplines that either plan, support, or deliver content. Her explorations led her to a taxonomy of content-related disciplines that includes more than 50 fields and areas of study and practice. Although we will analyze the many disciplines related to content later in this book, it is worth mentioning that Feldman proposes the following categories when talking about the multidisciplinary nature of content:

- Primary content disciplines that focus on methodology, delivery, and evolution of content itself (e.g., content design, content development, content engineering, content management, and content strategy)
- Secondary content disciplines that create content as a significant byproduct of their work (e.g., content curation, content marketing, conversation design, and customer experience)
- Tertiary content disciplines that are enabled by content (e.g., quality assurance, sales, engineering, and library science)

The primary content disciplines (as Feldman calls them)—those concerned with the separation of form and content and with workflows and technologies that treat content as computable data—impact in significant ways the roles and activities of technical communicators working with this kind of content.

In this book, we focus mainly on those primary "disciplines" and areas, but our analysis includes the categories presented by Feldman and others. Furthermore, we question if all content-related activities and tasks identified by Feldman can be labeled as disciplines. Instead, we propose elevating *content* as a discipline. Content possesses unique intellectual and institutional characteristics in industry and academia that warrant its treatment as a discipline.

And here's where we need to establish the specific area of the discipline of content where the work presented in this book resides. A few curious undergraduate students have walked into our college courses that include the word *content* in the title expecting readings and assignments related to the creation and marketing of digital content mainly aimed at social media platforms. The growing complexity of the discipline of content without a doubt includes academic and workplace conversations about digital creation for YouTube and social media. However, our work focuses on that specific type of content that, according to Rebekka, "has been given various names, including intelligent content, nimble content, smart content, portable content, and future-ready content" (Andersen, 2014, p. 116).

Ann Rockley, Charles Cooper, and Scott Abel (2015) describe intelligent content as "designed to be modular, structured, reusable, format free, and semantically rich and, as a consequence, discoverable, reconfigurable, and adaptable" (p. 1), and that is the content variant at the heart of this book—including all the names that it has received as it intersects with the discipline and practice of technical communication. In subsequent chapters, we refer to this type of content as *technical content*. We recognize that using the term *technical* may limit understandings to only content focused on technical subjects (what comes to mind when we think of technical information or technical documentation); however, we use the term *technical content* to refer to any type of content that is treated as computable data for the purpose of being both human and machine readable, adaptable, reusable, and portable. This can include other kinds of content, such as marketing and training content, which some organizations have integrated with product content in efforts to break down content silos and streamline content workflows.

Talking about content as its own discipline allows us to talk about the multidisciplinary interactions of content and technical communication and about the content industry, which Eichhorn (2022) has defined as "an umbrella term for all types of companies in the business of owning or providing digital content of any kind, from text publications to music to movies" (p. 7). The rise of the content industry, suggests Eichhorn, represents a "shift away from understanding cultural production in relation to aesthetic and material categories and toward understanding cultural production solely through a market-focused lens" (p. 27). This shift resonates with debates about the commodification of technical writing. From a writing studies perspective, Dush (2015) points out that content, as "assets" connected to "networked storage and technologies that require investment, and as provisionally valued items moving through networks," is "visible to market evaluation in a way that is not necessarily the case with writing bound to paper" (p. 177). The discipline of content treats content as a valuable, computable asset, and with that comes the need to examine opportunities and implications for the field of technical communication.

The Next Generation of Technical Content Professionals: A Research Project

Our quest to better understand the underexplored and not-well-understood activities and roles of technical communicators working on technical content teams began in 2017. We originally had the idea of hosting a workshop for leaders of technical communication and content-related activities in industry and academia to gather opinions and ideas about the future of the field(s). We had in mind an event similar to the 2000 Milwaukee Symposium hosted by Barbara Mirel and Rachel Spilka. That symposium brought together 18 technical communication specialists from industry and academia to discuss the status and future of the field. Whereas that symposium focused primarily on "research questions and issues that...participants deem[ed] crucial for the field to investigate in order to make greater progress in the coming decades" (Mirel & Spilka, 2002, pp. xv–xvi), the workshop that we had in mind would focus primarily on the following overarching research question:

> What can industry leaders, university administrators, educators, and recent graduates contribute to the development of curricula that prepare students to be leaders and innovators in the content-focused sector of technical communication?

Our hope was to identify concrete ideas and suggestions on how to update curricula at the assignment, course, and program level to help programs think through ways in which they could build pathways for students to pursue careers in the discipline of content.

We looked into securing funding for such a symposium from our respective universities as well as through a National Science Foundation grant. When these potential funding sources did not pan out, we decided to design a multi-staged project that allowed us to bring our questions to various stakeholders in the field. The IRB-approved project included two workshops at academic and professional conferences, a series of interviews with technical communication experts from industry and academia, a series of interviews with recent alumni of technical communication programs who now work as technical communicators, and a survey of managers of technical communication teams (Refer to Figure 1).

We conducted the first workshop at the 2017 ACM SIGDOC conference and the second workshop at the 2017 CIDM Best Practices Conference. Our primary goal was to document perspectives and suggestions on how technical communication curricula might be re-envisioned. We focused the first half of each workshop on the state of the profession of technical communication and invited participants to discuss in small groups what changes they were seeing in the profession, what skills entry-level technical communicators were expected to have coming into the profession, and what strengths and deficiencies new college graduates were bringing to the profession. We focused the second half of each workshop on the state of education in technical communication and again invited participants to discuss in small groups a series of questions. The questions this time focused on strengths and deficiencies in programs and curricula, skills our programs should be teaching and approaches for teaching them, and ideas for how our programs can best prepare students to be leaders and innovators in the changing profession. All large- and small-group discussions were recorded, transcribed, and analyzed.

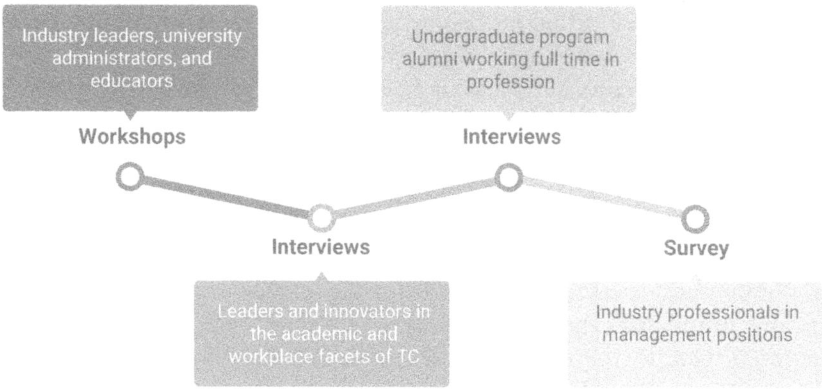

FIGURE 1 Research Stages of Next Generation of Technical Content Professionals Project

At the SIGDOC workshop, we had 28 participants, mostly from academia but several working in both industry and academia and several working solely in industry. At the Best Practices workshop, we had 15 participants, mostly from industry at the managerial level.

In 2018 and 2019, based on input collected at the workshops, we created a series of interview questions and conducted 31 semi-structured interviews with 17 industry experts in one or more disciplines related to technical content and 14 academic experts in programs that continually update curricula to align with industry practices. We identified these individuals as experts in technical communication based on their reputation in the field and on the extent to which they were considered thought leaders—whether that be through shaping best practices in the discipline of content, being widely cited and called upon for content-focused workshops and conference talks, or innovating as academic administrators or faculty by developing curricula responsive to industry changes. Seven of the industry experts had an active presence as educators or college-level instructors. We focused on equity and inclusion, as well, and actively pursued participants from traditionally underrepresented groups.

Collectively, interview participants helped us map the landscape of changing technical communicator roles and activities at both entry and advanced levels in organizations that have adopted workflows and technologies that treat technical content as computable data. They also helped us identify preparation needs for roles that interface with technical and product content. Our main questions were as follows:

- What does it mean to be a technical communicator today?
- How do you see technical communicator roles and activities changing? At the entry level? At the advanced level?

- How do you see hiring needs changing, particularly at the entry level and particularly for content-focused jobs?
- What do you see as implications of changing roles and hiring needs for education in technical communication?

In addition to interviewing industry and academic experts in the field, we wanted to gather perspectives of recent alumni of undergraduate programs in technical communication who had moved to full-time jobs as technical communicators. We wanted to learn about their experiences applying for jobs in the field and the extent to which their undergraduate training prepared them for particular content roles and activities in their current organizations. For this stage of the project, we reached out to program administrators and faculty as well as industry managers for names of recent alumni (within the past five years) who might be interested in participating in an interview. In 2019, 12 alumni from undergraduate programs offering a BA or BS in technical communication and one alumnus from a graduate MS program participated in semi-structured interviews. These alumni were working in a wide range of technical communication positions such as technical writer, technical specialist, technical trainer, technical editor, senior project consultant, senior e-learning analyst, and communications and engagement specialist.

Because only six of the 17 industry experts whom we had interviewed were in management roles and had extensive experience hiring for entry-level positions and onboarding new hires (the others primarily served in consultant roles), we knew that we needed to reach many more managers to get a clearer picture of how hiring needs were changing for content-focused jobs. In 2021, for the final data-collection stage of the project, then, we designed a 15-question survey aimed at managers and recruiters of technical communication professionals. The survey asked a series of questions focused on hiring needs and challenges, the skills and qualities managers and recruiters look for in new hires and why, the skills and qualities necessary for advancement to more strategic positions, and suggestions for how undergraduate programs might evolve to better prepare students for technical content-focused jobs. Several interview participants had discussed the growing need for young people in the field, especially as those who have been actively shaping best practices and leading teams retire; the problem, they suggested, is that most young people are not prepared to move into the roles that companies need to be filled. A key goal of conducting the survey was to better understand the preparation gap and identify strategies and approaches for closing the gap. We distributed the survey to an email listserv of information development managers and to eight relevant LinkedIn groups. Fifty-two self-identified managers or recruiters who met our criteria for inclusion completed the survey (we eliminated from analysis any survey responses from those not in manager or recruiter roles).

In Chapters 3, 4, and 6 of this book, we discuss findings from the survey and from our series of interviews with experts and alumni of undergraduate programs. We report on the ways in which technical communicator roles and activities are changing on technical content teams and how these changes have impacted hiring needs for both entry- and advanced-level positions, and we share research participants' views on the skills and competencies students need for today's technical content jobs and for leading and innovating in the content-focused sector of the field. As it has been the case with many ongoing research projects across academic disciplines, publications related to this project were slowed by challenges that we both had to navigate during the height of the COVID-19 pandemic. Further, at the time that we conducted our interviews and survey, generative AI tools were in their infancy and had yet to gain wide-scale public attention. Thus, neither our research questions nor our participants' responses focused on AI. Because of this, AI is not a dominant theme in this book. We do, however, describe in Chapter 2 how AI fits into the content professional's toolkit, and we highlight in Chapter 3 what leaders in the content and AI space have so far had to say about how activities and roles are changing now and will likely change in the future as AI is more fully integrated into content workflows. In Chapter 7, we provide some recommendations for content-related activities that can incorporate AI into technical communication curricula.

This book contributes to the wealth of existing research focused on training and education in technical communication (e.g., Baehr, 2015; Clegg et al., 2021; Kimball, 2015; Tham, Howard, & Verhulsdonck, 2022) and particularly to research that identifies the skills and competencies expected from graduates in the diverse and changing landscape of technical communication-related jobs (e.g., Batova & Andersen, 2017; Brumberger & Lauer, 2015; Henschel & Melonçon, 2014; Hovde & Renguette, 2017; Lanier, 2018; Shalamova, Rice-Bailey, & Wikoff, 2018; Stanton, 2017; Verhulsdonck, Howard, & Tham, 2021). Our field has known for a long time that technical communicators need to be more than excellent writers and communicators, that they need both conceptual and practical skills for knowledge work in the profession (Cargile Cook, 2002; Henschel & Melonçon, 2014). But our field does not yet have a good understanding of the skills and qualities needed for entry-level work and more strategic work in organizations that have adopted workflows and technologies that treat technical content as a valuable, computable asset. This book begins to identify these skills and qualities and to articulate their implications for undergraduate programs wanting to provide curricular pathways for students to pursue careers that connect technical communication to the discipline of content.

In this book, we aim to complement in particular the forward-looking education-focused work not only of Campbell and Swisher (2023); Tham, Howard, and Verhulsdonck (2022); Verhulsdonck, Howard, and Tham (2021); and Bridgeford (2020), but also that of Getto, Labriola, and Ruszkiewicz (2019, 2022) and Carlos, in his book *Creating Intelligent Content with Lightweight DITA* (Evia, 2018). Our chapters also connect to the work of Cleary (2021) in her book *The Profession and Practice of Technical Communication*, which reports stories and insights on practices that now define the larger profession of technical communication. We bring to these conversations the perspectives and experiences of dozens of experts and managers working in the discipline of content, and we intertwine these perspectives and experiences with those of recent program alumni now working as technical communicators, and of academic administrators and faculty known for developing curricula responsive to industry changes. We draw on these collective perspectives and experiences to map the landscape of changing technical communicator roles, activities, and preparation needs at both entry and advanced levels in organizations that treat technical content as computable data. Findings from workshops, interviews, and our survey point to opportunities for program and curricular innovations sensitive to different program strengths, existing program foci, and institutional constraints.

Book Contributions and Organization

This book contributes to ongoing conversations in the field regarding industry practices and education and training in technical communication in two important ways:

- It explores, theorizes, and analyzes the discipline of content, a transformative component of our field's collective identity. We offer a catalog of areas and disciplines that expands the vocabulary of what content is, does, and needs to be—that moves us away from lumping any discussion of content under the umbrellas of content management or content strategy. The catalog is useful for educators looking to develop curricula and student learning outcomes and for researchers looking to identify new areas of inquiry and a vocabulary for describing content-focused research.
- It presents empirical findings from a multi-staged, multi-year research project that examined changing activities, roles, and hiring needs and

identified the skills and qualities that hiring managers and key voices in the content industry say are most needed for entry-level technical content positions and for advancement to more strategic positions. This book thus contributes new knowledge to conversations around education and training in the field, particularly to conversations focused on aligning undergraduate programs with industry practice. We present recommendations from managers for program, curriculum, and course design, and we provide administrators and educators guidance on how they might innovate at the program level to better prepare students for content jobs and career mobility while also keeping an eye toward program sustainability. We additionally provide practitioners and managers strategies for partnering with university programs.

We do not aim to argue for a particular view of the field. Rather, we recognize the value and richness that ongoing debates bring to how we practice, research, and teach technical communication. This book aims to increase our collective understanding of the discipline of content and of the skills and qualities that our students might develop to meaningfully contribute human-centered design values and critical sensibilities to the content ecosystems of their future organizations.

This book is divided into three sections: Discrepancies, Challenges, and Opportunities; Roles, Activities, and Career Trajectories; and Competencies, Curricular Pathways, and Programmatic Innovations.

Section I: Discrepancies, Challenges, and Opportunities

In this section, we demonstrate growing employment needs for technical communicators who are prepared to take on varied content roles in organizations that have adopted workflows and technologies that treat content as computable data. We present the discipline of content as a framework for understanding changing roles, activities, and hiring needs in these organizations and for understanding implications of these changes for education and training in the field of technical communication.

- Chapter 1 argues for making room for the discipline of content in the field of technical communication's collective identity. Given the field's diverse spheres of practice and diverse areas of research and research approaches, scholars have struggled to pinpoint a common thread that connects the different facets of the field. In this chapter, we refrain from further problematizing the lack of a common thread and instead align our project with Melonçon and Schreiber's (2022) call to embrace the field's collective identity, comprising various components that are either becoming, stabilizing, or re-stabilizing. The discipline of content, we argue, is a "becoming" component in need of extensive study and theorizing, as academic researchers and educators in the field are only just beginning to understand the landscape of modern content operations that treat (mainly written) content as computable data. This chapter reviews debates around definitions of the field and connections between research, education, and practice and advocates for increased attention in our educational programs and our research to the discipline of content, now a key facet of our field's collective identity.

- Chapter 2 is dedicated to introducing the different areas and disciplines that either plan, support, or deliver technical content. It describes the specific areas and practices of content strategy, content design, content engineering, and content operations, which have emerged to address the who, what, when, why, how, and where of technical content that needs to be orchestrated for a multitude of devices and delivery channels. We place technical content creation—the primary activity of entry-level content-focused technical communicators—in the disciplinary area of content design; doing so gives us a way to talk not only about the entry-level position but also about more advanced positions in other content-related areas. By defining four specific areas that comprise the discipline of content, we expand the vocabulary of what content is, does, and needs to be; these four areas serve as a framework for understanding changing roles, activities, and hiring needs and their implications for education and training in technical communication. The need to define the discipline of content and its related areas, we argue, is a direct response to the growing adoption of digital transformations in organizations, the foundation of which is structured content. The chapter ends with a discussion of the implications of digital transformations for content, including the growing impact of AI.

Section II: Roles, Activities, and Career Trajectories

In this section, we discuss findings from workshops and interviews with leaders in industry and academia, from more interviews with recent alumni of technical communication programs, and from a survey of hiring managers. We report on the ways in which roles and activities are changing on technical content teams and how these changes have impacted hiring needs for both entry- and advanced-level positions. We also share research participants' views on the skills and qualities students need for technical content jobs and for leading and innovating in the discipline of content.

- Chapter 3 reports on the ways in which entry-level technical communicator roles and activities are changing in organizations that are undergoing or have undergone digital content transformations. Structured authoring and Agile development practices have had the greatest impact on technical communicator roles and activities at all levels; the diverse products and content outputs that organizations now produce and the software development processes as well as the tools and platforms that support content activities are key drivers of change. Chapter 3 describes the what, why, and how of structured authoring and Agile development practices and demonstrates, through our reporting of interview and survey findings, the growing need for technical communicators as content professionals. Chapter 3 also highlights the perspectives of five content professionals who had recently completed undergraduate degrees in technical communication; these professionals describe their current roles and responsibilities, the publishing environments of their organizations, and the skills and qualities needed to move into more advanced content-focused positions. The chapter ends with a description of the impact of generative AI on technical content processes and what this means for preparing technical content professionals.

- Chapter 4 calls attention to the aging workforce of technical content professionals and to strategy conversations among many in the content industry on how to attract and prepare a new generation of professionals. Drawing on interview and survey findings, we describe the kinds of positions for which managers need to hire and the essential functions of those positions. We then report on the skills and qualities that hiring managers and key voices in the content industry identified as most needed for entry-level content-focused positions and for advancement to more strategic positions. These skills and qualities include technical knowledge and fluency, a learning mindset, structured authoring, and strong writing and communication skills. We report, as well, on the varied challenges that these leaders have faced when hiring for particular roles and skill sets and on the strengths and deficiencies that they have observed new college graduates bringing to their organizations and to the larger content profession. We further share the job search stories of recent alumni of undergraduate programs in technical communication and related fields who are now working full time as technical communicators.

Section III: Competencies, Curricular Pathways, and Programmatic Innovations

Drawing on our research findings, this section extends existing literacy frameworks in technical communication to include knowledge work in the discipline of content and, in doing so, introduces a competencies framework for content professionals. It further provides administrators and educators guidance on how they might develop new or enhance existing curricular pathways that prepare students as technical content professionals. A key challenge in revising any program is the need to navigate institutional constraints and to consider regional employment needs. Given the diversity of our field and the many constraints under which programs operate, this section emphasizes different kinds of pathways that different kinds of programs might adopt.

- Chapter 5 highlights arguments for keeping education programs connected to professional practice and summarizes how technical communication scholars have identified and described needed competencies and skills, particularly those that connect with practice, for students completing undergraduate degrees in technical and professional communication. Approaches to identifying competencies and skills have included updates to prior literacy frameworks based on examinations of workplace practices; analyses of job advertisements and interview and survey reports; and, for competencies and skills specific to technical content work, syntheses of the content-focused practitioner discourse focused on roles, activities, and desired skill sets. We offer detailed lists of the competencies and skills identified through these approaches and make the case for developing a competencies framework specific to work in the content discipline. Such a framework would identify the competencies and qualities that college graduates need to both gain access to and then succeed and advance in entry-level content-focused positions.
- Chapter 6 situates the skills, competencies, and qualities needed for content professionals within the literacy frameworks discussed in Chapter 5. We specifically expand Henschel and Melonçon's (2014) explanatory matrix of conceptual and practical skills to map with the competencies entry-level content professionals need to not only land content-focused positions but also meaningfully contribute to the content initiatives of their organizations—as critical pragmatic practitioners. In expanding the matrix, we introduce a competencies framework for content professionals that we hope administrators and educators will find useful for the purpose of planning programs and curricular innovations. The framework identifies seven conceptual skills for content professionals: systems knowledge and thinking, rhetorical proficiency, skills in abstraction and research, social proficiency, an experimentation mindset, and technological literacy. For each conceptual skill, it also lists associated activities, approaches, technologies, and/or mindsets with which content professionals must engage. The chapter ends with a discussion of curricular gaps and opportunities for innovation in US undergraduate programs in technical communication.

- Chapter 7 introduces ideas for situating the intersection of technical communication and the discipline of content in college curricula. Given the limitations that undergraduate courses and programs in technical communication and related fields face at different institutions, we are careful to avoid generalizations and absolute recommendations. For programs interested in building pathways for students to pursue careers in the discipline of content, we suggest a staged approach, with each stage increasing in complexity and engagement; for each stage, we describe topics to cover and why and offer both resources and suggested activities. Chapter 7 additionally shares the advice that both industry and academic leaders had for technical communication educators on how they could keep curricula aligned with the needs of the content-focused facet of the field. The number one recommendation was to emphasize structured authoring throughout curricula. Other key recommendations included prioritizing industry outreach and partnerships, forming interdisciplinary alliances and integrating interdisciplinarity into curricular pathways, and engaging in professional development.

References

Abel, S. (2014). Content. In S. Abel & R. Bailie (Eds.), *The language of content strategy* (pp. 12–13). Laguna Hills, CA: XML Press.

Andersen, R. (2014). Rhetorical work in the age of content management: Implications for the field of technical communication. *Journal of Business and Technical Communication*, *28*(2), 115–157.

Andersen, R., & Batova, T. (2015). The current state of component content management: An integrative literature review. *IEEE Transaction on Professional Communication*, *58*(3), 247–270.

Atherton, M. & Hane, C. (2018). *Designing connected content: Plan and model digital products for today and tomorrow*. San Francisco, CA: New Riders.

Baehr, C. (2015). Complexities in hybridization: Professional identities and relationships in technical communication. *Technical Communication*, *62*(2), 104–117.

Batova, T., & Andersen, R. (2017). A systematic literature review of changes in roles/skills in component content management environments and implications for education. *Technical Communication Quarterly*, *26*(2), 1–28.

Bridgeford, T. (2020). *Teaching content management in technical and professional communication*. New York: Routledge.

Brumberger, E., & Lauer, C. (2015). The evolution of technical communication: An analysis of industry job postings. *Technical Communication*, *62*(4), 224–243.

Campbell, K., & Swisher, V. (2023). A maturity model for content strategy development and technical communicator leadership. *Journal of Technical Writing and Communication*, *53*(4), 286–309.

Cargile Cook, K. (2002). Layered literacies: A theoretical frame for technical communication pedagogy. *Technical Communication Quarterly*, *11*(1), 5–29.

Cleary, Y. (2021). *The profession and practice of technical communication*. New York: Routledge.

Clegg, G., Lauer, J., Phelps, J., & Melonçon, L. (2021). Programmatic outcomes in undergraduate technical and professional communication programs. *Technical Communication Quarterly*, *30*(1), 19–33.

Dush, L. (2015). When writing becomes content. *College Composition and Communication*, *67*(2), 173–196.

Eichhorn, K. (2022). *Content*. Cambridge: The MIT Press.

Evia, C. (2018). *Creating intelligent content with Lightweight DITA*. New York: Routledge.

Feldman, S. (2020, April 11). A list of content disciplines. Retrieved April 15, 2024, from http://www.sarafeldman.com/effective-content/a-list-of-content-disciplines/.

Getto, G., Labriola, J., & Ruszkiewicz, S. (2019). *Content strategy in technical communication*. New York: Routledge.

Getto, G., Labriola, J., & Ruszkiewicz, S. (2022). *Content strategy: A how-to guide*. New York: Routledge.

Gollner, J. (2016, December 28). Content 4.0 [Blog post]. Retrieved June 26, 2019, from https://www.gollner.ca/2016/12/content_4-0.html.

Halvorson, K., & Rach, M. (2012). *Content strategy for the web* (2nd ed.). Indianapolis, IN: New Riders.

Henschel, S., & Melonçon, L. (2014). Of horseman and layered literacies: Assessment instruments for aligning technical and professional communication undergraduate curricula with professional expectations. *Programmatic Perspectives*, *6*(1), 3–26.

Hovde, M. R., & Renguette, C. C. (2017). Technological literacy: A framework for teaching technical communication software tools. *Technical Communication Quarterly*, *26*(4), 395–411.

Kimball, M. (2015). Training and education: Technical communication managers speak out. *Technical Communication*, *62*(2), 135–145.

Kimball, M. (2016). The golden age of technical communication. *Journal of Technical Writing and Communication*, *47*(3), 330–358.

Lanier, C. (2018). Toward understanding important workplace issues for technical communicators. *Technical communication*, *65*(1), 66–84.

Masycheff, A. (2018). Keeping technical writing jobs relevant with content automation technologies. LinkedIn. Retrieved June 10, 2022 from https://www.linkedin.com/pulse/keeping-technical-writing-jobs-relevant-content-alex-masycheff/.

Melonçon, L., & Henschel, S. M. (2013). Current state of U.S. undergraduate degree programs in technical and professional communication. *Technical Communication*, *60*(1), 45–64.

Melonçon, L., & Schreiber, J. (2022). Introduction: Promoting a sustainable collective identity for technical and professional communication. In J. Schreiber & L. Melonçon (Eds.), *Assembling critical components: A framework for sustaining technical and professional communication* (pp. 3–16). WAC Clearinghouse: University Press of Colorado.

Mirel, B., & Spilka, R. (2002). Introduction. In B. Mirel & R. Spilka (Eds.), *Reshaping Reshaping*ogram and c *technical communication: New directions and challenges for the 21st century* (pp. 1–6). Mahwah, NJ: Lawrence Erlbaum Associates.

Molisani, J., & Abel, S. (2012, February). Tech comm 2.0: Reinventing our relevance in the 2000s. *Intercom*, 14–18.

Rockley, A., Cooper, C., & Abel, S. (2015). *Intelligent content: A primer*. Laguna Hills, CA: XML Press.

Shalamova, N., Rice-Bailey, T., & Wikoff, K. (2018). Evolving skill sets and job pathways of technical communicators. *Communication Design Quarterly*, *6*(3), 14–24.

Simple[A] (2019, June 30). The future of content starts with collaboration. Retrieved January 17, 2025, from https://simplea.com/Articles/future-of-content-collaboration.

Stanton, R. (2017). Do technical/professional writing (TPW) programs offer what students need for their start in the workplace? A comparison of requirements in program curricula and job ads in industry. *Technical Communication*, *64*(3), 223–236.

Swarts, J. (2010). Recycled writing: Assembling actor networks from reusable content. *Journal of Business and Technical Communication*, *24*(1), 127–163.

Technically Write IT (2018, May 30). Information 4.0- Making content intelligent. Retrieved January 17, 2025, from https://technicallywriteit.com/information-4-0/.

Tham, J., Howard, T., & Verhulsdonck, G. (2022). Extending design thinking, content strategy, and artificial intelligence into technical communication and user experience design programs: Further pedagogical implications. *Journal of Technical Writing and Communication*, *52*(4), 428–451.

Verhulsdonck, G., Howard, T., & Tham, J. (2021). Investigating the impact of design thinking, content strategy, and artificial intelligence: A "Streams" approach for technical communication and user experience. *Journal of Technical Writing and Communication*, *51*(4), 468–492.

Virtaluoto, J., Sannino, A., & Engeström, Y. (2016). Surviving outsourcing and offshoring: Technical communication professionals in search of a future. *Journal of Business and Technical Communication*, *30*(4), 495–532.

1

EMBRACING CONTENT IN TECHNICAL COMMUNICATION'S COLLECTIVE IDENTITY

A Collective Identities Approach to Technical Communication

The field of technical communication, including related and offspring fields such as professional communication, science communication, medical writing, technical editing, user experience (UX), social media strategy, usability studies, entrepreneurial communication, and more, is expansive and diverse. As Melonçon and Schreiber (2022) write, technical (and professional) communication "now addresses a range of industries, organizations, sites, and locations including everything from technology to healthcare to nonprofits," and it presents a wide array of career opportunities spanning all facets of communication, and represents a growing number of certificates as well as graduate and undergraduate programs (p. 3).

The all-encompassing field of technical communication allows for the study, practice, and teaching of functional communication, writing, and information design in all of its forms in academic, industry, government, and non-profit contexts. It also provides academic institutions a suitable label to distinguish it from sister academic disciplines like mass communication and marketing. Technical communication—at least as defined in academic contexts in the United States—often includes the more narrow but still vast facet that focuses specifically on the communication of technical or specialized content.

Diversity in Practice

In the field of technical communication, practitioners and scholars can be specialists in writing, communication, or translation. They can be instructional

DOI: 10.4324/9781003451730-3

designers or information architects. They can work in content organizations as content managers, content strategists, and more. They can also work on UX teams—being designers, evaluators, or creators of UX. In their analysis of close to 1,000 US information development positions posted to Monster.com, Eva Brumberger and Claire Lauer (2015) found "enormous variety in position titles," with the majority of them falling into the categories of content developer/manager or technical writer/editor. Results of a global Adobe Technical Communication Survey (Singh, 2018) of over 2,000 technical communicators also revealed diverse job titles, with respondents identifying as technical writers, technical documentation managers, UI/UX designers, information developers, and architects, among other titles. And even more diverse than job titles in the field are the myriad roles in which technical communicators now work—roles often not reflected in a single position title (see, e.g., Shalamova, Rice-Bailey, & Witkoff, 2018).

In a 2018 census of the state of the technical communication industry, Saul Carliner and Yuan Chen (2018a) found that a majority of technical communicators work in technology (24%), information technology (IT) services and solutions (15%), and manufacturing (8%) industries, with the following industries having about 5% of the population of technical communicators: healthcare IT, professional services, government and military, finance and banking, and engineering, architecture, and construction (p. 11). In addition to this range of industries, technical communicators work in a variety of internal departments, including, but not limited to, IT and information services, research and development, manufacturing, and marketing and sales (p. 10). In these industries and departments, technical communicators engage in various activities (e.g., developing content, managing projects and groups, planning strategies) and produce diverse communication products (e.g., user guides, help and user assistance topics, reference material, policies, and procedures) (2018b, pp. 14–16). As Carliner and Chen (2018b) note, the census gives a broad picture of the work of technical communicators, but it is limited in that "it does not provide deep insights into the unique characteristics of the jobs of each individual technical communicator" (p. 16).

Diversity in Research

In addition to and partly a result of its diverse spheres of practice, technical communication comprises diverse areas of research approached through a wide range of research perspectives (e.g., social, critical, theoretical, empirical) and methods (e.g., qualitative, quantitative, historical). A review of recent book titles in the ATTW Series in Technical and Professional Communication from Routledge demonstrates the wide spectrum of research interests and areas that comprise the field; these include, for example, environmental risk, the profession and practice of technical communication, design thinking, content strategy, translation and localization, editing, social media, citizenship and advocacy, project management, sustainable program innovation, scientific and medical communication, and plain language.

In mapping the research questions most examined in technical communication, Carolyn Rude (2009) identified four broad areas into which our field's research tends to fall. These include disciplinarity (e.g., What are our definitions, history, status, possible future, and research methods?), pedagogy (e.g., What should be the content of our courses and curriculum?), practice (e.g., What design practices are inclusive of international users and users with disabilities?), and social change (e.g., How do texts function as agents of knowledge making, action, and change?). These areas of research remain essential in an ever-evolving and changing field, and the questions that emerge within them continue to require different research approaches. Research examining questions in the area of social change has particularly taken off since the publication of *Technical Communication after the Social Justice Turn* (Walton, Moore, & Jones, 2019). Equity, justice, and inclusion are now central concerns of a growing body of technical communication-related scholarship that examines how systems and structures facilitate injustice and that explores approaches for redressing injustice and enacting social justice in the varied contexts in which technical communicators work and the varied communications that they help design. The 2022 launch of the new *Journal of Technical Communication and Social Justice*, as well as several special journal issues focused on social justice (e.g., March 2022 issue of the *IEEE Transactions on Professional Communication*), firmly solidifies social justice as a core area of academic research in the field.

Research focused on disciplinarity, pedagogy, and practice also continues to expand. Recent disciplinarity research has addressed questions concerning how we define the field (e.g., Cleary, 2021; Henning & Bemer, 2016; Jones, Moore, & Walton, 2016; Melonçon & Schreiber, 2022; St.Amant & Melonçon, 2016) and what we research in the field (e.g., Andersen & Hackos, 2021; Boettger & Friess, 2020; Lam, 2021). Research focused on pedagogical and curricular approaches to preparing students for technical communication work has particularly grown, with bodies of pedagogical scholarship developing around different facets of the field, such as social justice (e.g., Agboka & Dorpenyo, 2022; Walton & Agboka, 2021), UX (Getto, Labriola, & Lancaster, 2023; Rose & Turner, 2020; Tham, Howard, & Verhulsdonck, 2022), intelligent content (Evia, 2018) and content management (Bridgeford, 2020a), and international and intercultural communication (e.g., Batova, 2018; St.Amant, 2017). This scholarship informs programmatic design and curricular innovations in important ways, and to an extent, has shaped the many different foci and strengths of undergraduate programs.

Research focused on modern technical communication practice, while still one of the four key areas of research identified by Rude (2009), has been less robust in the last ten years (see recent analyses of topic patterns in Boettger & Friess, 2020; Carradini, 2022; Lam, 2021). Though the topics that technical communicators research are ever-expanding given how technological innovations shape modern practice (e.g., augmented reality, artificial intelligence [AI]), the amount of research on those topics remains small relative to the amount of research focused on the more foundational topics of rhetoric and genre (Boettger & Friess, 2020).

The Common Thread Problem

With so much diversity in what technical communicators do, what they call themselves, where they work, and what they research, scholars in the field have struggled to pinpoint a common thread that connects the different facets of the field and helps us explain—collectively—our value to different stakeholders outside of the field. Kirk St.Amant and Lisa Melonçon (2016) suggest that the field's diverse, multifaceted nature has led to a problem of incommensurability—"a situation where nothing seems shared or common" (p. 269). Different disciplinary perspectives, different theoretical approaches, different educational backgrounds and pathways, and different program flavors and institutional locations all feed into the diversity of curricula that define our programs and the diversity of topics we research, including the methods we use for conducting our research. The problem of incommensurability, St.Amant and Melonçon argue, makes who we are and what we do not easily recognizable and "undermines our legitimacy as a discipline and our power to act, engage, and develop as a field" (p. 270). They suggest that our field's research might be the best mechanism for revealing what disparate parts of the field have in common, and, as a way to unify the field, they encourage its members to identify research questions that cut across its disparate parts.

Another proposed solution to the problem of incommensurability comes from Teresa Henning and Amanda Bemer (2016), who call for enlarging the *Occupational Outlook Handbook*'s definition of technical communicator—which currently focuses on the deliverables that technical communicators produce and disseminate through various organizational channels. The definition does not articulate the value that technical communicators bring to their organizations or the range of strategic problem-solving activities they perform; the consequence, the authors suggest, is what Saul Carliner (2012) called a lack of "brand identity," or an identity that "clarifies for others what services technical communicators provide" (Henning & Bemer, 2016). Henning and Bemer thus advocate for an enlarged definition that clearly articulates what technical communicators can do and that would "grant the field power in presenting a united front to employers with respect to the value technical communicators bring to the workplace" (p. 2).

We suggest that creating such a definition risks either overpromising what a technical communicator can do or not being specific enough to make clear what value *diverse* kinds of technical communicators bring to different contexts. This is specifically challenging when we move to the facet of the field that drives our work in this book: the intersection of technical communication and the discipline of content.

For example, some technical communicators bring UX design and writing expertise to product teams, whereas others bring business analysis and data mining expertise to content strategy teams. One definition cannot encompass all the activities that now facilitate the creation, management, and delivery of content across various channels and platforms. One definition cannot encompass the contributions of an information architect versus a technical writer versus a content engineer working in publishing environments ranging from web-only to print documents-only to multichannel publishing.

Henning and Bemer (2016) suggest that "clearly defining what a technical communicator has the ability to do can be a means for giving technical communicators the power to be influential in a particular context, and when appropriate, the capacity to direct others" (p. 12). We agree, but we caution that any enlarged definition of technical communication should not promise a span of abilities that some technical communicators have and others do not, given the vast range of experiences and educational backgrounds from which the diverse universe of technical communicators comes. What would be more productive for all, we propose, is not to work toward one definition that condenses the entire field, but rather a collection of definitions that explain—and advocate for—different components or facets of the field, with the discipline of content being one of them.

A Collective Identities Approach

Lisa Melonçon and Joanna Schreiber (2022) advocate for embracing the field's diversity in their introductory chapter to the collection, *Assembling Critical Components: A Framework for Sustaining Technical and Professional Communication*. Recognizing the range of interests, perspectives, and stakeholders as both a strength and a challenge for the field, Melonçon and Schreiber push against calls for an enlarged definition that tries to encompass and unite all of its facets or a more narrow definition that would exclude some perspectives and practices in favor of others. They instead advocate for embracing the field's collective identity, an approach that focuses on the "component parts that make up this collective identity" and that celebrates the "knowledges and practices" that comprise these parts (p. 4). When we think of the field as having different identities, we can better align with a specific group that shares practices, qualities, values, expectations, and more. A different identities approach, the authors suggest, "both creates a space for new knowledges and approaches as well as establishes a 'collective identity' that moves us beyond fragmentation so that the various aspects of the field may work and grow together" (p. 4).

Melonçon and Schreiber draw on Gilles Deleuze and Felix Guattari's (1987) concept of assemblage in explaining the affordances of a collective identity approach. Assemblage as an ontological framework allows space for an "ongoing state of becoming or coming into being"; this state stabilizes and re-stabilizes "as the field shifts and changes, offering a body of knowledge from which to draw definitions and make claims" (p. 7). When technical communication is viewed as a field comprising various components that are either becoming, stabilizing, or re-stabilizing, we can imagine how technical communicators whose primary job is to design content automation workflows (say, the "discipline of content" component of the field) can exist alongside technical communicators whose primary job is to build more just and inclusive voting policies and protocols (say, the "social justice" component of the field). We can imagine how expertise in the discipline of content may inform the ways in which policies and protocols are designed for goals such as multichannel publishing or translation, and we can imagine how expertise in social justice may (and should) inform the development of content models, taxonomies, and other forms of infrastructural writing that shape digital transformations. The foundational knowledges and practices of different facets of the field overlap and diverge in ways that allow the components to achieve the varied goals and outcomes of the institutions, non-institutions, and stakeholders served.

Allowing room for "becoming" or "coming into being" makes the collective identity approach particularly appealing to us. When new components or facets are brought into this collective identity, they require extensive study and theorizing—a process of sensemaking that begins to build a body of knowledge. At first, new components may comprise a set of evolving practices that are not reflected in our research or curricula; as a body of knowledge establishes and matures, it can inform, better define, and refine those practices and guide curricular decisions for programs interested in creating space for the new component. Melonçon and Schreiber (2022) point to the field's embrace of issues of social justice as one example of a "becoming" component that is now central to the field's assembled identity in US-based academic circles. Lisa Dush (2015) argued for a similar embracing of the becoming of "content," that which "writing has become" (p. 174).

We view the discipline of content as a "becoming" component in need of extensive study and theorizing, and perhaps more urgently than other becoming components given the changes in the strategic space of technical communication that it represents. Certainly, the field already has a small but robust body of scholarship on the shifts to modular content development, single sourcing, and the separation of form and content (e.g., Andersen & Batova, 2015; Bridgeford, 2020a; Campbell & Swisher, 2023; Evia, 2018; Gattis, 2023; Swarts, 2022). But academic researchers and educators in the field have yet to engage with modern operations that optimize content production and leverage content as a valuable asset, and we don't yet have a way of describing the different, essential content activities that comprise what many in industry call the disciplines or trades of content. These practices now extend far beyond the boundaries/realm of product documentation in a growing number of organizations; modular technical content is now served up in marketing materials, training videos, medical equipment interfaces, customer service bots powered by large language models, and more, and it has become an important part of brand building—of pre-sales and post-sales information products that all need to deliver a uniform message. At the core of our field's collective identity—the through-line connecting all of the components—is human-centered design. We have much to lose if we don't bring to the forefront of our identity the discipline of content, which defines through areas such as strategy, design, and engineering how humans can and should interact with and experience content. And we have much to lose if we do not create curricular space in our programs that prepares students for careers as strategists, architects, and other important roles in organizations that have adopted workflows and technologies that treat content as a valuable, computable asset.

Barbara Mirel and Rachel Spilka (2002) acknowledged nearly 25 years ago that the move away from documents to content modules represented a substantial change in the composing of communications. They expressed concern then that "content and structure are divided between content strategists and information architects and integrated by yet other information designers, often threatening cohesion, coherence, and unity that make communications effective" (p. 1). The discipline of content has emerged in part as a response to the need to better coordinate content operations in organizations—in fact, content operations is its own emerging disciplinary area (see Evia, 2024). Over the years, content strategy and content engineering experts have addressed through documented best practices many of the original concerns about communications comprising content modules, but we still have little research on the practices or theories and approaches underlying the discipline of content. And we do not yet understand well how roles and activities have changed in these environments or what skills and qualities these roles and activities require.

Challenges and Opportunities

The discipline of content is a "becoming" facet of technical communication, and we need to better understand it and develop a better, more consistent way to talk about it. In this section, we turn to exploring why this facet of technical communication's collective identity has been underexplored and remains not well understood.

Difficulty Staying Abreast of Content Trends

In the "Afterword: Beyond Management" that we authored for Tracy Bridgeford's (2020a) collection, *Teaching Content Management in Technical and Professional Communication*, we described how "keeping abreast of trends—and, more importantly, making sense of what they mean for educating future technical communicators—is proving to be an increasingly difficult task" (Evia & Andersen, 2020, p. 214). The dearth of research in the field on content activities in organizations seems to reflect this difficulty, and we would not be surprised if many researchers shy away from studying content because, as we wrote in the Afterword, "talking about content in technical communication is complicated, and continues to get more complicated" (p. 214). The concept of Content 4.0 is a good example. This next evolution of content (now a reality for many organizations) has been characterized as follows:

> (C)ontent is planned, designed, created, managed, and exchanged as objects that incorporate … the associated rules governing the structure and meaning of the content and an array of rendition and behaviour processes that the object can use to render the material independently or in concert with other objects.
>
> *(Gollner, 2016)*

The content activities described in Gollner's definition look very different from the content activities described in traditional document development environments or even single-sourcing environments where content is created once and reused in multiple deliverables.

Staying abreast of how these processes work and the contributions and expertise required to make them work requires one to be connected to industry and have access to the space and time required to not only conduct empirical research in/on these environments but also to understand and learn the languages, standards, and technologies that support these processes (for those who do not come from a computer science or engineering background, this learning curve can be steep). For many academics, contributing to the body of scholarship or developing expertise in this facet of the field feels daunting at best and impossible at worst, given lack of professional development time, pressures of tenure and merit/promotion clocks, and absence of rewards for focusing research on the discipline of content over other facets of the field comprising our collective identity. And so, we attribute the scant attention to technical content in our field's body of research and in our program curricula in part to the complexities around following and embracing content trends and talking about content (for discussions on scant attention in research, refer to Boettger & Friess, 2020 and Lam, 2021; for discussions on scant attention in our program curricula, refer to Batova & Andersen, 2017 and Bridgeford, 2020b).

Declining Research on Practice-Focused Topics

Some conversations on the academic side of technical communication have been focused on concerns about pragmatism, or problem-solving and efficiency. These concerns are present in discussions of educational goals and curricular priorities (see, e.g., Bernhardt, 2002; Kimball, 2015). Recent studies indicate that technical communication research focused on practice in the sphere of what Kimball (2016) called "strategic" technical communication, or that which happens within formal organizations, has diminished, but this trend is not new. Carolyn Rude (2015) suggested that the decline of research focused on improving practice began in the 1990s when researchers sought to establish technical communication as an academic discipline recognized for contributions to human knowledge. Ryan Boettger and Erin Friess (2020) found that the content areas researchers were studying had not changed in 30 years. The areas generating the most publications included rhetoric, genre, pedagogy, and diversity (with the journals *Technical Communication* and *IEEE Transactions on Professional Communication* being the exceptions). More practice-focused topics such as usability/UX, knowledge and information management, design, and editing and style did not appear in their sample as frequently as expected (p. 18). Because the foundations of the field have evolved, and because technical communicators now work in all phases of product development, not just the final stage, Boettger and Friess had expected to see their sample reflect this evolution.

In two different studies, Chris Lam (2021) and Erin Friess and Ryan Boettger (2021) found little overlap between the content produced in practitioner versus academic forums, further indicating that academic researchers were not necessarily engaging in topics important to practitioners and to the practice of technical communication in formal organizations. Lam (2021) demonstrated this disconnect in his analysis of the #TechComm community on the social media platform then known as Twitter. He found that practitioners and academics who actively engaged in the community were having very different conversations, with practitioners primarily focusing on "documentation and the tools and technologies related to software documentation" (p. 17) and academics primarily focusing on content relevant to other academics. Lam also found that the topic of DITA/XML dominated practitioner conversations spanning a four-year period—an unsurprising finding given the widespread adoption of the DITA standard[1] by content organizations across industries. In our own review of the academic literature, we found very few studies that had specifically examined DITA as an authoring platform, and they were either by Carlos and his frequent collaborator Michael Priestley (Evia & Priestley, 2016) or by our colleague Jason Swarts (2022). Numerous articles and books had recognized the prominence of DITA in technical communication practice, but DITA itself had not been a topic of study in those publications.

Friess and Boettger (2021), too, demonstrated a disconnect between content produced in practitioner versus academic forums. They compared content published in *Intercom* (the magazine of the former Society for Technical Communication) versus the field's peer-reviewed journals from 1996 to 2017 and found that professionalization and technology were central topics in *Intercom*, whereas rhetoric, diversity, and pedagogy were central topics in the peer-reviewed journals (p. 13). The stark difference in topics presented what they saw as "a fractured picture of TC" that indicated a lack of shared interests, priorities, and goals between practitioners and academic researchers.

The relatively low percentage of empirical research published in the field also suggests that practice-focused research in the sphere of strategic technical communication has not been a priority research area for some time, due likely to reasons already mentioned but also to barriers to conducting empirical research in formal organizations. Lisa Melonçon and Kirk St.Amant (2019), in their analysis of empirical research published in the field's peer-reviewed journals between 2011 and 2015, identified 119 out of 321 articles that used empirical research methods (this number did not include classroom- or pedagogy-based studies). This means that only 37% of the articles published in the five flagship journals during a four-year period drew on empirical methods. Boettger and Friess (2020) speculated that most research on rhetoric, genre, pedagogy, and diversity—the topics generating the most publications in the field's peer-reviewed journals, was not empirical. They pointed to a study by Chris Lam and Ryan Boettger (2017) that found that "68.8% of rhetoric, 75% of pedagogy, and 53.8% of genre research was non-empirical" (p. 20). This and other findings pointing to research-practice disconnects indicate that the field could benefit from more empirical research focused on practice in formal organizations.

We have seen in the past few years several journal special issues focused on content strategy (Getto, Flanagan, & Labriola, 2023), UX (Getto, Labriola, & Lancaster, 2023; Rose & Schreiber, 2021), and AI (Carradini, 2024, 2025), and these issues have included a handful of empirical studies. The increased attention to these emerging, more practical facets of the field suggests that many scholars also recognize the need for more research that reflects, informs, and helps shape the practice of technical communication. We hope that this book additionally inspires research that helps scholars better understand the emerging discipline of content.

In their Editorial for a special issue of *Technical Communication*, Chris Dayley and Isidore Dorpenyo (2023) acknowledge that "little research has been done regarding the state of diversity in the professional practice of technical and professional communication in the U.S. and across the globe" (p. 1). We agree with them and consider that the discipline of content could particularly benefit from research that investigates content practices from a social justice perspective, with the goal of informing practice and enacting change. We have, in fact, been encouraged by recent research that examines AI ethics and machine learning from a social justice perspective and that connects those practices to the roles of practicing content professionals (e.g., Duin & Pedersen, 2023; Hocutt, Ranade, & Verhulsdonck, 2022). We have also been encouraged by what appears to be a shift in practice in the discipline of content toward diversity, equity, inclusion, and social justice. Particularly after the murder of George Floyd in 2020, colleagues in industry working on content-related sectors have talked to us about the impact of structural racism and inequalities in their professions. Jonathan McFadden (2024), Senior Content Designer at Shopify, writes that the same tide of introspection that took over social media and mainstream communication channels in 2020 "swept over the content community, sparking a fever pitch of conversations, presentations, books, and blogs about inclusive content and methods for expunging racist, ableist, ageist, sexist, homophobic, and other harmful language, components, and biases from products their companies created" (pp. 150–151).

In her book *From Solo to Scaled: Building a Sustainable Content Strategy Practice*, Natalie Marie Dunbar (2022) also asks for intentional approaches to inclusion and diversity when building content strategy teams, with a focus on the diversity of consumers of content while also arguing for more diversity among those developing the content. We have seen, as well, an increased number of conference talks and webinars focused on strategies for centering users in content models, taxonomies, and designs with the aim of providing users with seamless, consistent, equitable, accessible, and memorable content experiences. There is clear interest among content professionals to better understand ways in which social justice can inform content practices.

Many technical communication researchers over the years have articulated compelling reasons for our field to stay connected with practice. Some have argued that because our field is a practice-oriented discipline, our research and educational programs must necessarily focus on practice and strive to influence practice (see, e.g., Albers, 2016; Davis, 2013; de Jong, 2009; Rude, 2009, 2015). Aligning technical communication with medicine and engineering, Davis (2013) insisted that, like those fields, our field "must be immersed in solving real-world problems" (p. 30). Rude (2009), too, emphasized parallels between technical communication and other practice-oriented disciplines and suggested that the knowledge we produce through research "must be recognized as enabling better work and use of the products it helps present to users" (p. 205). Because our disciplinary identity is tied to practice, these researchers see the field's primary aim as that of solving real-world problems and supporting practitioners in their jobs; and because practitioners' identities are largely defined by the work they do, researchers and educators must necessarily keep a finger on the pulse of industry—on the practices that define what Kimball (2016) termed the glass age of technical communication. Most recently, Jason Swarts (2024), reflecting on the rise of content operations as a game-changing field of practice and study, emphasized the need for new basic and applied research that examines content operations practices and that leads to education—curricular evolution (p. 160). In Chapter 2, we introduce content operations as a key area of the discipline of content focused on the management, or governance, of all content activities.

A Space for the Discipline of Content in Our Collective Identity

Human advocacy has always been at the heart of technical communication work—in corporations, government and non-profit organizations, educational institutions, and communities. It is what distinguishes technical communication from other fields concerned with technology design and use. Kathy Pringle and Sean Williams (2005) articulated human advocacy as a value set that defines what we do and for whom:

> We approach technology from a human perspective and believe that technology should adapt to people, not the other way around. We design

> our communication products accordingly, using whatever media, software, technology, or tool is most appropriate to achieve this end. People, we would argue, are the ultimate end, not the technology. Not only is this a value set in technical communication, it's also the way we work. We use tools not as an end but as a means to help people.
>
> *(p. 369)*

These values represent the field's long commitment to human-centered communication design. These are the values that technical communicators have historically enacted as rhetorical agents overseeing the processes by which texts (or "communication products") are created, managed, and delivered to those who need them. In technical communication, pragmatism and human-centered design are interconnected, and it is the human perspective—the focus on human contexts of use, human goals, needs, and expectations—that technical communicators have historically contributed to problem-solving activities in organizations.

In organizations that have adopted workflows and technologies that treat (mainly written) content as computable data, text-making activities are governed not by writers-as-rhetorical agents but by a web of content technologies, standards, and codified processes designed by business leaders, software developers, content strategists, information architects, and more, some of whom may have little or no training in the humanities. While technical communicators in advanced positions often assume these strategic roles, industry survey data indicates that most technical communicators working at this level did not study technical communication in an academic institution. As Rebekka has argued before, those who have primary rhetorical agency in these content ecosystems "are, for the most part, those who design the system" that "governs what content can be written and how it must be written, structured, and tagged to support reuse and dynamic publishing" (Andersen, 2014). The nature and location of rhetorical work thus look very different in these contexts than in more traditional document development contexts.

What is at stake when humanities-trained technical communicators are not key contributors to the design and iterative improvement of content initiatives? What is at stake if our educational programs are not preparing students to meaningfully contribute human-centered design values and critical sensibilities to the different components comprising content initiatives, from content strategy to taxonomy development to content modeling and the generation of content by AI engines? When the texts that users rely on every day to make critical decisions, perform actions, and solve problems are the output of highly technical, codified systems and processes, those of us who have been trained in human-centered design approaches to communication must necessarily ask critical questions and act to ensure that these texts help, not hurt, those for whom they are intended. Technical communicators who are prepared to assume strategic roles in content organizations are needed more than ever, which means we as educators need to understand what skills and qualities employers expect for these roles so that we can develop curricula that ensure our students are viable candidates for entry-level positions that afford them opportunities to advance.

While Kimball (2016) may be right that there are fewer jobs for the "smallish group of people paid well to think strategically about design, rhetoric, visualization, presentation, information architecture, technology, and usability" than there are for the "larger, less-well-paid group" of people charged with writing text for the "content machine" (p. 10), we take the position that the field has a moral imperative to prepare students for strategic roles; there may be fewer positions, but that should not mean that programs should focus curricula entirely on other components of our collective identity. Too much is at stake if we fail to embrace the discipline of content as the significant "becoming" component that it is. We thus advocate for creating curricular pathways that prepare students for work in one or more areas of the discipline of content. Creating such pathways does not mean programs need to re-invent themselves, abandon existing strengths, or hire loads of new faculty; rather, such pathways can be additive and can leverage the strengths of other fields that have a stake in content, such as design, business management, marketing, product management, computer science, and software engineering.

Nearly 25 years ago, Mirel and Spilka (2002) articulated new directions for the field given the network of activities contributing to the development of technological and communication products, from business analysis to planning to research and design. They recognized the need for technical communicators to expand their "vision, goals, and concerns to new categories" beyond the communication products that they produced and implored practitioners "to take on boundary-spanning roles organizationally and become the originators and purveyors of information and knowledge crucial for strategic planning and decisions about product directions" (p. 91). "Only through this larger role," they argued, "can [technical communicators] significantly affect technological usefulness, usability, learnability, enjoyment, and quality" (p. 92). The same argument holds true today in content organizations; only through boundary-spanning roles that interface with strategic planning, design, and execution can technical communicators help ensure that ethical considerations; diversity, equity, and inclusion; UX; human beings; and other field priorities are centered in decision-making. Curricular pathways that prepare students for entry-level content-focused jobs will need to focus on these boundary-spanning roles and the expansive skills and qualities they require. Doing so will afford students opportunities to move into strategic roles and to lead and innovate in those roles.

Calls for curricular evolution in the content-focused facet of our field's collective identity are certainly not new (see Batova & Andersen, 2017; Bridgeford, 2020b; Evia, 2018; Getto, Labriola, & Ruszkiewicz, 2019; Swarts, 2020, 2024; Tham, Howard, & Verhulsdonck, 2022; Verhulsdonck, Howard, & Tham, 2021). For example, in her introduction to *Teaching Content Management in Technical and Professional Communication*, Tracy Bridgeford (2020b) called for undergraduate programs to embrace content management as "a pedagogical exigency" and noted that more than ten years after George Pullman and Baotong Gu (2008) urged the field to integrate content management into curricula, "our pedagogical community has not responded with any sense of urgency" (p. 1). Chapter authors in Bridgeford's (2020a) edited collection from 2020 respond to this exigence by articulating varied approaches to teaching content management at the course level. Jason Tham, Tharon Howard, and Gustav Verhulsdonck (2022), also responding to this sense of urgency, identified content strategy, AI, and design thinking as three streams that are "reshaping mainstream TPC (technical and professional communication) practices" (p. 3) and making "next-level programmatic innovation" a necessity for preparing "TPC students for a UX and AI-driven world" (p. 5). The authors, moving beyond just calling for curricular change, articulate strategies for teaching these streams in technical and professional communication courses. Notably, at the crux of these streams is the discipline that defines new ways of creating, managing, and delivering content.

Note

1 We will focus on the DITA standard for content in Chapter 3.

References

Agboka, G. Y., & Dorpenyo, I. K. (2022). Curricular efforts in technical communication after the social justice turn. *Journal of Business and Technical Communication*, *36*(1), 38–70.

Albers, M. (2016). Improving research communication. *Technical Communication*, *63*(4), 293–297.

Andersen, R. (2014). Rhetorical work in the age of content management: Implications for the field of technical communication. *Journal of Business and Technical Communication*, *28*(2), 115–157.

Andersen, R., & Batova, T. (2015). The current state of component content management: An integrative literature review. *IEEE Transaction on Professional Communication*, *58*(3), 247–270.

Andersen, R., & Hackos, J. (2021). Practicing technical communicators' experiences with and perspectives on academic publishing. *Technical Communication*, *68*(3), 29–55.

Batova, T. (2018). Global technical communication in 7.5 weeks online: Combining industry and academic perspectives. *IEEE Transactions on Professional Communication*, *61*(3), 311–329.

Batova, T., & Andersen, R. (2017). A systematic literature review of changes in roles/skills in component content management environments and implications for education. *Technical Communication Quarterly*, *26*(2), 1–28.

Bernhardt, S. (2002). Active practice: Connecting academia and industry in technical and professional communication. In B. Mirel & R. Spilka (Eds.), *Reshaping technical communication: New directions and challenges for the 21st century* (pp. 81–90). Mahwah, NJ: Lawrence Erlbaum Associates.

Boettger, R., & Friess, E. (2020). Content and authorship patterns in technical communication journals (1996–2017): A quantitative content analysis. *Technical Communication*, *67*(3), 4–24.

Bridgeford, T. (2020a). *Teaching content management in technical and professional communication*. New York: Routledge.

Bridgeford, T. (2020b). Introduction: Content management: A pedagogical exigency. In T. Bridgeford (Ed.), *Teaching content management in technical and professional communication* (pp. 1–15). New York: Routledge.

Brumberger, E., & Lauer, C. (2015). The evolution of technical communication: An analysis of industry job postings. *Technical Communication*, *62*(4), 224–243.

Campbell, K., & Swisher, V. (2023). A maturity model for content strategy development and technical communicator leadership. *Journal of Technical Writing and Communication*, *53*(4), 286–309.

Carliner, S. (2012). The three approaches to professionalization in technical communication. *Technical Communication*, *59*(1), 49–65.

Carliner, S., & Chen, Y. (2018a). Who technical communicators are: A summary of demographics, backgrounds, and employment. *Intercom*, 8–12.

Carliner, S., & Chen, Y. (2018b). What technical communicators do. *Intercom*, 13–16.

Carradini, S. (2022). The ship of Theseus: Change over time in topics of technical communication research abstracts. In J. Schreiber & L. Melonçon (Eds.), *Assembling critical components: A framework for sustaining technical and professional communication* (pp. 39–67). WAC Clearinghouse: University Press of Colorado.

Carradini, S. (Ed.). (2024). Effects of artificial intelligence tools in technical communication pedagogy, practice, and research, part 1. [Special issue]. *Journal of Business and Technical Communication*, *38*(3).

Carradini, S. (Ed.). (2025). Effects of artificial intelligence tools in technical communication pedagogy, practice, and research, Part 2 [Special issue]. *Journal of Business and Technical Communication*, *39*(1).

Cleary, Y. (2021). *The profession and practice of technical communication*. New York: Routledge.

Davis, M. (2013). Identifying core principles and expectations. *Communication Design Quarterly*, *1*(4), 29–30.

Dayley, C., & Dorpenyo, I. (2023). Practices, reflections, and methodologies: What is successful diversity, equity, and inclusion in the technical communication workplace? *Technical Communication*, *70*(3), 1–5.

de Jong, M. (2009). Practically meaningful academic research and academically relevant practitioner research. *Technical Communication*, *56*(2), 97–98.

Deleuze, G., & Guattari, F. (1987). *A thousand plateaus: Capitalism and schizophrenia* (B. Massumi, Trans.). Minneapolis, MN: University of Minnesota Press.

Duin, A. H., & Pedersen, I. (2023). *Augmentation technologies and artificial intelligence in technical communication: Designing ethical futures*. New York: Routledge.

Dunbar, N. M. (2022). *From solo to scaled: Building sustainable content strategy practice*. New York: Rosenfeld Media.

Dush, L. (2015). When writing becomes content. *College Composition and Communication*, *67*(2), 173–196.

Evia, C. (2018). *Creating intelligent content with Lightweight DITA*. New York: Routledge.

Evia, C. (2024). *Content operations from start to scale: Perspectives from industry experts*. Blacksburg: Virginia Tech Publishing.

Evia, C., & Andersen, R. (2020). Afterword: Beyond management: Understanding the many forces that shape content today. In T. Bridgeford (Ed.), *Teaching content management in technical and professional communication* (pp. 213–231). New York: Routledge.

Evia, C., & Priestley, M. (2016). Structured authoring without XML: Evaluating Lightweight DITA for technical documentation. *Technical Communication*, *63*(1), 23–37.

Friess, E., & Boettger, R. K. (2021). Identifying commonalities and divergences between technical communication scholarly and trade publications (1996–2017). *Journal of Business and Technical Communication*, *35*(4), 407–432.

Gattis, L. (2023). A study of lexical repetition and the comprehensibility of single-sourced technical documents. *Journal of Technical Writing and Communication*, *53*(4), 335–355.

Getto, G., Flanagan, S., & Labriola, J. (Eds.). (2023). The people, practices, and technologies central to content strategy [Special issue]. *Journal of Technical Writing and Communication*, *53*(4).

Getto, G., Labriola, J., & Lancaster, A. (2023). The state of UX in technical and professional communication: Courses, programs, and jobs. *IEEE Transactions on Professional Communication*, *66*(4), 317–321.

Getto, G., Labriola, J., & Ruszkiewicz, S. (2019). *Content strategy in technical communication*. New York: Routledge.

Gollner, J. (2016, December 28). Content 4.0 [Blog post]. Retrieved June 26, 2019, from https://www.gollner.ca/2016/12/content_4-0.html.

Henning, T., & Bemer, A. (2016). Reconsidering power and legitimacy in technical communication: A case for enlarging the definition of technical communicator. *Journal of Technical Writing and Communication*, *46*(3), 311–341.

Hocutt, D., Ranade, N., & Verhulsdonck, G. (2022). Localizing content: The roles of technical & professional communicators and machine learning in personalized chatbot responses. *Technical Communication*, *69*(4), 114–131.
Jones, N., Moore, K., & Walton, R. (2016). Disrupting the past to disrupt the future: An antenarrative of technical communication. *Technical Communication Quarterly*, *25*(4), 211–229.
Kimball, M. (2015). Training and education: Technical communication managers speak out. *Technical Communication*, *62*(2), 135–145.
Kimball, M. (2016). The golden age of technical communication. *Journal of Technical Writing and Communication*, *47*(3), 330–358.
Lam, C. (2021). Hashtag #TechComm: An overview of members, networks, and themes from 2016–2019. *Technical Communication*, *68*(2), 5–21.
Lam, C., & Boettger, R. (2017). An overview of research methods in technical communication journals (2012–2016). In *2017 IEEE International Professional Communication Conference (ProComm)* (pp. 1–4).
McFadden, J. (2024). Epilogue: You're here because of people. In C. Evia (Ed.), *Content operations from start to scale: Perspectives from industry experts* (pp. 147–154). Blacksburg: Virginia Tech Publishing.
Melonçon, L., & Schreiber, J. (2022). Introduction: Promoting a sustainable collective identity for technical and professional communication. In J. Schreiber & L. Melonçon (Eds.), *Assembling critical components: A framework for sustaining technical and professional communication* (pp. 3–16). WAC Clearinghouse: University Press of Colorado.
Melonçon, L., & St.Amant, K. (2019). Empirical research in technical and professional communication: A 5-year examination of research methods and a call for research sustainability. *Journal of Technical Writing and Communication*, *49*(2), 128–155.
Mirel, B., & Spilka, R. (2002). Introduction. In B. Mirel & R. Spilka (Eds.), *Reshaping Reshaping technical communication: New directions and challenges for the 21st century* (pp. 1–6). Mahwah, NJ: Lawrence Erlbaum Associates.
Pringle, K., & Williams, S. (2005). The future is the past: Has technical communication arrived as a profession? *Technical Communication*, *52*(3), 361–370.
Pullman, G., & Gu, B. (2008). Guest editors' introduction: Rationalizing and rhetoricizing content management. *Technical Communication Quarterly*, *17*, 1–9.
Rose, E., & Schreiber, J. (Eds.). (2021). User experience and technical communication: Beyond intertwining [Special issue]. *Journal of Technical Writing and Communication*, *51*(4).
Rose, E., & Turner, H. (2020). The pedagogy of user experience: Methods, tools, and approaches. In *Proceedings of the 38th ACM International Conference on Design of Communication* (pp. 1–2).
Rude, C. D. (2009). Mapping the research questions in technical communication. *Journal of Business and Technical Communication*, *23*(2), 174–215.
Rude, C. D. (2015). Building identity and community through research. *Journal of Technical Writing and Communication*, *45*(4), 366–380.
Singh, A. (2018, March 21). Adobe tech comm survey 2017–2018 [Webinar]. Retrieved June 26, 2019, from https://techcomm-trends-2018.meetus.adobeevents.com/.
Shalamova, N., Rice-Bailey, T., & Wikoff, K. (2018). Evolving skill sets and job pathways of technical communicators. *Communication Design Quarterly*, *6*(3), 14–24.

St.Amant, K. (2017). Of friction points and infrastructures: Rethinking the dynamics of offering online education in technical communication in global contexts. *Technical Communication Quarterly*, *26*(3), 223–241.

St.Amant, K., & Melonçon, L. (2016). Addressing the incommensurable: A research-based perspective for considering issues of power and legitimacy in the field. *Journal of Technical Writing and Communication*, *46*(3), 267–283.

Swarts, J. (2020). Writing about structure and DITA. In T. Bridgeford (Ed.), *Teaching content management in technical and professional communication* (pp. 155–175). New York: Routledge.

Swarts, J. (2022). Signaling context in topic-based writing. *Technical Communication*, *69*(1), 40–53.

Swarts, J. (2024). Afterword. In C. Evia (Ed.), *Content operations from start to scale: Perspectives from industry experts* (pp. 155–162). Blacksburg: Virginia Tech Publishing.

Tham, J., Howard, T., & Verhulsdonck, G. (2022). Extending design thinking, content strategy, and artificial intelligence into technical communication and user experience design programs: Further pedagogical implications. *Journal of Technical Writing and Communication*, *52*(2), 428–459.

Verhulsdonck, G., Howard, T., & Tham, J. (2021). Investigating the impact of design thinking, content strategy, and artificial intelligence: A "Streams" approach for technical communication and user experience. *Journal of Technical Writing and Communication*, *51*(4), 468–492.

Walton, R., & Agboka, G. (2021). *Equipping technical communicators for social justice work: Theories, methodologies, pedagogies*. Logan, UT: Utah State University Press.

Walton, R., Moore, K., & Jones, N. (2019). *Technical communication after the social justice turn*. New York: Routledge.

2

EMERGENCE OF THE DISCIPLINE OF CONTENT

The Intersection of Technical Communication and Content Strategy

In August 2007, Carlos was preparing his syllabi for the two courses that he was assigned to teach for the undergraduate major in Professional and Technical Writing at Virginia Tech: ENGL 3814: *Creating User Documentation* and ENGL 4814: *Writing for the Web*. He had been teaching both courses regularly for almost three years, and although they had clear differences, they also shared some foundational fabric that at the time was difficult to describe using disciplinary terms from technical communication. *Creating User Documentation* was known among the students as "the DITA course," and *Writing for the Web* was known as "the HTML course" or simply "the web course." Both courses had readings, assignments, and activities related to audience analysis, design and development of structure, layout, textual and multimedia information, and methods for evaluating if (and how) the information products created were solving problems for the intended audiences. During the syllabus preparation process, Carlos found a web article by Rachel Lovinger (2007) that gave him new language to describe those courses and, eventually, his career as a technical communicator. Titled "Content Strategy: The Philosophy of Data," Lovinger's article attempted to define the "new" practice of content strategy. In her article, Lovinger announced the following:

> Everything is content? What about design? Yes, it's content. Structure? Content. Metadata? Also content. You probably expected a more incisive analysis than that. Well, how about, "Literally, everything is content."
>
> *(2007)*

DOI: 10.4324/9781003451730-4

Although the technical communication literature had mentioned content strategy prior to 2007, it was presented as related to technical documentation. For example, CJ Walker (2023) found "(a)necdotal evidence that content strategy [was] being used as a term among early content professionals" back in 1997. But in 2007 the Lovinger article revealed to Carlos that the thread connecting "the DITA course" to "the web course" was content—more specifically, the nascent *discipline* of content strategy. As we write this chapter, Carlos is preparing to teach a section of the old "web course," which now has been rebranded as "Developing Intelligent Content."

At the same time in 2007, some 720 miles northwest of the Virginia Tech campus, Rebekka was immersed in the research sites of her doctoral dissertation. As a PhD student in English at the University of Wisconsin-Milwaukee, Rebekka was studying how (then) new practices in technical communication like automation for content management and reuse were affecting the dynamics of work groups in industry. Lovinger's article also provided language that Rebekka's work, placed as well in that crossroads of technical communication and content strategy, needed—even if she was not aware of it back then. Lovinger emphasized content challenges and solutions like the following:

> As we start to design and build websites with massively larger volumes of content, we find that often they've outgrown the ability of individuals to manually organize them. Now we need automation and complex algorithms to find that needle in the haystack. We need the content to include inherent meaning that makes sense to machines, for example, to support data-driven applications based on search, browse, and related links.
>
> *(2007)*

For Carlos and Rebekka, and many others in our field around 2007, technical communication was an established discipline (with the disciplinary challenges that we mentioned in the previous chapter, but a discipline notwithstanding), and content strategy was just a practice that probably involved using spreadsheets to count content assets.

Then came Kristina Halvorson in 2008 to proclaim that content strategy was to be regarded as a discipline. She exhorted her readers to "(t)ake up the torch for content strategy. Learn it. Practice it. Promote it." Commenting on Lovinger's (2007) call for content strategists to be "experts in all aspects of communication," Halvorson proposed instead to embrace "a number of content-related disciplines that deserve their own definition." Halvorson's (2008) preliminary list of content-related disciplines included the following:

- Editorial strategy
- Web writing

- Metadata strategy
- Search engine optimization
- Content management strategy
- Content channel distribution strategy.

In a blog post titled "Ten Years Later: The Discipline of Content Strategy," Halvorson (2018) identified Ann Rockley's (2003) *Managing Enterprise Content* as an influential book on her original proclamation of the discipline of content strategy. With its chapters on structured authoring and DITA, reuse, and single sourcing, Rockley's book covered some of the topics that Carlos and Rebekka were working on in 2007. As a consequence, we can place Halvorson's discussion about content-related disciplines in that same alley that connects technical communication to content strategy.

Even back in 2008, Halvorson was not only talking about content strategy as its own discipline but also including other content-related disciplines. The blurry disciplinary lines around content are extended in the taxonomy of content-related disciplines compiled by Sara Feldman (2020), which we mentioned in the "Introduction" to this book. The multidisciplinary nature of content, therefore, warrants our proposal for elevating *content* itself as a discipline. Additionally, as we covered in the previous chapter, we view the discipline of content as a "becoming" component of the discipline of technical communication.

But with so much talk about the discipline of this and the discipline of that, the next section provides a brief overview of what we mean by treating content as a discipline (or *the* discipline, as opposed to treating other content-related areas as disciplines) and how this label frames our view of content-related activities and roles in the world of technical communication. Then, we will propose where to place entry-level technical communication professionals in the discipline of content and how to position them as contributors to efforts related to digital transformations in organizations that have adopted workflows and technologies that treat content as computable data. Digital transformation is generally understood as the process of transforming or re-imagining an organization through the adoption of new or updating of existing technologies and tools. We discuss, as well, how AI is being integrated into content workflows and how it fits into the content professional's toolkit.

Why Treat Content as *the* Discipline?

When we started working on the proposal that eventually became this book, we struggled with our use of the term "discipline" in the book's title. After all, Carlos had been recently appointed Director of the Academy of Transdisciplinary Studies at Virginia Tech, and in presentations for the 2019 Center for Information-Development Management IDEAS

Conference and the 2019 Society for Technical Communication Conference, we had emphasized the multidisciplinary nature of content as a big mixing of writing studies, communication, computer science, and many other fields. Isn't the idea of disciplines, then, inaccurate to describe this becoming facet of our field? We acknowledge that the term "discipline" conveys many connotations with a complicated history—Roger Deacon (2002), for example, provides a detailed analysis of Michel Foucault's engagement with this term—and complicated present and future. However, we find the term "discipline" extremely helpful as a dynamic container for bodies of knowledge and practice related to the planning, development, and evaluation of digital content. We also favor this term to reflect its widespread adoption and usage in the academic literature and its popularity with industry voices (e.g., Barker, 2007; Collins, 2021). While we initially debated using the term "profession" to depict the content practices of professionals working in industry (the practices on which this book focuses), we realized that profession, with its ties to occupational identity and the workplace (see Savage, 2023), did not encompass the expertise, extensive body of knowledge, and validated strategies and approaches that now inform those practices, specifically those focused on strategy, design, operations, and engineering.

The discipline of content is a complex and active holder for many (and growing) disciplinary areas that, for us, reflects the now-famous observation made by Michael Crow, President of Arizona State University as of this writing, about academic disciplines:

> Knowledge knows no boundaries. The core disciplines are but one element of our intellectual identity. The traditional disciplinary organization of universities may not be the optimal way to organize knowledge, or to organize the institution itself, or to teach students, or to solve the social, economic, and technological challenges confronting institutions in the regions in which they are located. Accordingly, I encourage teaching and research that is interdisciplinary, multidisciplinary, and transdisciplinary. I encourage the convergence of disciplines, where appropriate, a practice that might more accurately be described as intellectual fusion.
>
> *(Crow, 2002, p. 26)*

Paradoxically, as Cyrille Rigolot (2020) suggests, transdisciplinarity can be treated as a new discipline and as a way of being. We, thus, perceive the discipline of content as an intellectual fusion of the transdisciplinary epistemologies and practices of content in academic and industry circles. We are not alone in our adoption of this term as a fusion of conversations about content. John Collins, Senior Content Architect and Content Engineer at Atlassian, authored a now oft-cited blog post titled "The maturing content discipline," which started with his expression of excitement over the growing maturity of content as a discipline.

Collins (2021) identifies the following four disciplinary areas (he calls them roles) in his version of the content discipline:

- Content design
- Content strategy
- Content operations
- Content engineering.

Collins's list shares some keywords and actions with the preliminary list of content-related disciplines introduced by Halvorson (2008) and is more manageable and detailed than Feldman's (2020) crowdsourced taxonomy of content-related disciplines—with more than 50 fields and areas of study and practice—which we discussed in the "Introduction" of this book. Collins's "roles" appear in the category of primary content disciplines (those that focus on methodology, delivery, and evolution of content itself) in Feldman's taxonomy.

In all fairness, the areas fused in different versions of the discipline of content are labeled, defined, and described depending on the brand and strategy that a consultant or industry leader is promoting. Some consultants and leaders do not distinguish between content-related disciplinary areas, opting to put all activities and roles under the umbrellas of content strategy (e.g., Casey, 2023) or content design (e.g., Atherton & Hane, 2018; McConnell, 2022; Winters, 2017). In technical communication academic circles, we have tended to cover everything with the banner of content management (e.g., Bridgeford, 2020; Pullman & Gu, 2008) and sometimes content strategy (e.g., Getto, Labriola, & Ruskiewicz, 2023). And when we talk about content strategy, this can mean anything from a spreadsheet of posts and pages for a single WordPress website to an enterprise-level, unified plan.

Others have made distinctions among the areas but have different views on which area owns or oversees particular activities. For example, there's a general consensus among industry leaders that content strategy, building on the early definition offered by Halvorson (2008), is focused on understanding business and customer goals and then, based on that understanding, planning for the creation, distribution, and governance of content (Casey, 2023; Halvorson, 2008; Bailie & Urbina, 2013). But what that means in terms of the essential activities that compose the disciplinary area varies. Among many other activities, Meghan Casey (2023) suggests that strategy includes specifying content structure and requirements and building a content model. Simple[A] (2016) suggests that strategy is more limited—it articulates the plan and requirements for content while content engineering oversees the technical implementation of the plan and requirements. Content engineering, according to Simple[A], defines and builds out the content structure, content semantics, content technology, metadata, taxonomy, and other enablers of "content processes such as authoring, validation, transformation, and publishing." Yet others, such as Mike Atherton and Carrie Hane (2018) and Rachel McConnell (2022), describe these essential activities as falling under the purview of content design, which they describe in much the same way as others have described content strategy but with an eye more toward UX design. Content operations, (or ContentOps, to use the term that's become part of the technology vernacular), probably sounds like a new entry in this fusion of disciplinary areas. Rahel Bailie (2024) counters that, "(i)n fact, Content Operations has been around as long as content has been produced as part of an organizational process" (p. 13) and offers examples of its usage in articles and presentations from the early 2000s.

As Halvorson proclaimed in 2008, content strategy can be regarded as a discipline on its own. In fact, if we follow the general characteristics of a discipline proposed by Fred D'Agostino (2012), content strategy checks some of the boxes. D'Agostino posits that disciplines are at the same time institutional and intellectual forms, and they support and are supported by elements including the following (condensed and paraphrased from his original list).

As an institutional form:

- a professional association which sponsors
    - conferences
    - awards and prizes
    - scholarly journals
    - a code of professional ethics
- an academic organizational unit in universities
- an undergraduate curriculum
- "canonical types of texts," including scholarly journals and research monographs
- undergraduate and coursework postgraduate teaching
- social networks.

As an intellectual form:

- criteria for assessing the value of work
- a concept of rigor
- a rhetoric that identifies its user as a member of the disciplinary community

- an accepted narrative of the discipline's development and legitimacy
- a clear agenda for research, including definitions of problems and of solutions
- a characteristic disciplinary methodology
- a body of literature
- a discursive community with a common language, including field-specific jargon
- rhetorical devices for displaying competence (based on D'Agostino, 2012, pp. 334–336).

Unquestionably, content strategy and other content-related areas or roles identified by Collins (2021) possess *some* institutional and intellectual disciplinary elements. However, as a collective under the term of the discipline of content, those areas can cover much more ground in their path to establishing themselves as a discipline—particularly when they intersect with an established discipline such as marketing, computer science, or technical communication.

The following section puts the spotlight on how those content-related disciplinary areas have been represented in academic and industry conversations to consolidate the discipline of content as a becoming component of our home discipline of technical communication.

Content Strategy

In the field of technical communication, Ann Rockley (2003) is a pioneer in explorations into content strategy, which she described as "a repeatable method of identifying all content requirements up front, creating consistently structured content for reuse, managing that content in a definitive source, and assembling content on demand to meet your customers' needs" (p. 12). Rockley then identified the following as key elements of a unified content strategy:

- A content management system to manage your content in a definitive source
- Reusable content objects that enable you to write content once and use it many times
- Unified processes that encourage people to work collaboratively, which results in processes that are repeatable and transparent, regardless of department or author.

Of the areas fused in the discipline of content, content strategy is the most developed and documented. We mentioned earlier in this chapter that several authors have followed Halvorson's (2008) definition of content strategy and have built some level of consensus about this area focusing on understanding business and customer goals and then, based on that understanding, planning for the creation, distribution, and governance of content. Several books from Halvorson and Rach (2012), Kissane (2011), Casey (2023), McGrane (2012), Wachter-Boettcher (2012), and Bailie and Urbina (2013), among others, have documented conversations and tactics about content strategy as it relates to many facets of digital content.

However, there are different interpretations of the activities that someone with "content strategist" as their title is expected to perform. Rahel Bailie (2019) offers the following analogy to explain the different "flavors" of content strategy:

> I often use the field of medicine to explain the range of specialties in content strategy: There are many types of doctors that practice in a variety of specialties, but they all have a basic medical degree as the common denominator. In content strategy, the common denominator is the systematization of content—creating an operational system that allows content to be produced reliably, at scale, and without degradation of quality.
>
> *(p. 121)*

Even Halvorson in recent years has moved away from mere website content strategy to define her work as focused on enterprise content strategy. In a 2023 episode of her podcast "The Content Strategy Podcast" (Halvorson, 2023), she discussed the shift as follows:

> ... it's become pretty clear to me that there are content, still not our best favorite word... but that enterprise content strategy is where we're really working to, whether or not it's centralize operations, at least come up with shared principles, shared structures, shared process, shared role understanding and definition so that content can become organized and synchronized and purposeful across an organization.

We concur with Halvorson and focus on content strategy (particularly when it is combined with technical communication) as it affects enterprise content or, as we said in the "Introduction" to this book, intelligent content. We acknowledge, however, that content strategy is bigger than these intersections, and that practitioners all over the world see this area primarily as connected to marketing and social media—as documented in the work of authors like Andy Crestodina (2022) and Robert Rose (2023).

Zooming in on the combination of content strategy and technical communication, the pioneer work from Rockley (2003) resonates in more recent academic-oriented publications, including three special journal issues dedicated to content strategy (Bailie, 2019; Batova &

Andersen, 2016; Getto, Flanagan, & Labriola, 2023) and two articles focused on intersections between content strategy, AI, and design thinking and implications for programmatic innovations (Tham, Howard, & Verhulsdonck, 2022; Verhulsdonck, Howard, & Tham, 2021). Rockley's work particularly resonates in publications from the team of Guiseppe Getto, Jack Labriola, and Sheryl Ruszkiewicz, who have contributed to this area mainly with an edited collection, *Content Strategy in Technical Communication* (2019), and a more how-to book, *Content Strategy: A How-to Guide* (2023), both published by the Association of Teachers of Technical Writing book series. Getto, Labriola, and Ruszkiewicz define content strategy as "a wide-ranging field that encompasses everything from people who manage websites for a living to people who create repositories of content that companies can use for documentation" (2023, p. xi). They include the following parts of the content strategy process:

- Audience analysis
- Identifying content types and channels
- Content auditing
- Assembling a content strategy plan
- Collaborating with other content developers
- Content modeling
- Revising and editing genres
- Ensuring content usability and accessibility
- Delivering, governing, and maintenance
- Localizing content
- Using tools and technologies (p. 18).

Although Getto, Labriola, and Ruszkiewicz mention collaboration with content developers, they don't include the specific task of content creation, which in recent years—at least in some prominent work sectors of the United Kingdom and United States—has been described as content design.

Content Design

As a concept for describing the processes of planning and producing content, content design is a rather recent import from the United Kingdom influenced by the work of Sarah Winters (2017). In her introductory book titled *Content Design*, Winters describes this area as follows:

> Content design means not limiting yourself to just words. Content on the web is often words, but not always. The point of content design is

> that you start with research to help you identify what your users actually need (which isn't the same as what they say they want).
>
> *(p. 2)*

In the United States, the proliferation of job ads and titles including "content design" started to replace more established terms including "copywriter," "technical writer," and even "content strategist" in some sectors. In fact, the UX Writing Hub (Keshtcher, 2022), in its annual salary survey report of writers in tech, reported a 13% increase between 2021 and 2022 in the number of survey respondents who had the title of content designer; at the same time, the Hub reported an 11% decrease in the number who had the title of technical writer or content strategist. Many attribute this change to the adoption of "content design" as a job process by Spotify and Facebook (see, e.g., Long, 2020).

In an episode of "The Content Strategy Podcast," Halvorson (2023) addressed the relationship among content design, UX, and Facebook:

> And then we come around to content design, which from the minute Facebook launched, from the minute any software-as-a-service launched online, UX writing content design was really a thing because what we've seen is that it's really product companies that have rallied around that title because content designers and UX writers are really working so actively and so closely on feature and experience design with designers and engineers.

Elisabeth Carr (2020) announced the decision to rebrand the Facebook Product Content Strategy team as the Content Design team in a now-famous post on the Design at Meta blog. Carr acknowledged that "content strategy is used to describe many flavors of content work: social media marketing, branded content creation, information architecture and the more UX-specific in-product design work that we do at Facebook." She added that, as the title of content strategist continues to broaden, her team conducted analysis of titles and roles "at 27 tech companies," and "we found that content design is rapidly becoming the standard for how people talk about the type of product content work we do at Facebook."

Carr then defined content designers as "people who design in words, concepts, systems and terminology, voice and tone, and who know how much these things matter in solving problems for the people who use our products around the world."

If these descriptions of content design sound similar to those of content strategy, it's because the two areas, as we have said earlier, tend to be labeled, defined, and described depending on the brand and strategy promoted by content leaders in industry (and some in academia, including Getto, Labriola, & Ruszkiewicz, 2023). Some leaders put the activities associated with content strategy under the umbrella of content design while others do the opposite, and many leaders tend not to make a distinction between content strategy and content design, instead choosing the label that works best for their or their companies' purposes. However, we find a distinction to be useful in that it allows for talking about the roles and activities associated with content development, and it offers those wanting to learn about the content discipline a productive way to distinguish the two disciplinary areas. We find Collins' (2021) distinction particularly applicable, as he links content strategy to planning and content design to creation. Content design, he writes, focuses on "creating or curating the content that meets [a] user need," which "could be words, but it also could be images, infographics, video, code, or even audio." Content design, then, is part of the process of implementing the content strategy plan.

Another factor contributing to the adoption of content design as a term and title by companies in the information technology sector is its connection (in name and in practice) to user experience (UX) design. Sarah Winters (2017) announced that "(t)here's a lot of technical skill in content design" (p. 6), and many of her recommendations are based on UX research practices (including user stories and content critiques, among others). More recent publications have expanded that bridge between content design and UX. Erica Jorgensen (2023), for example, defines content design as "the art and science of determining what specific content should appear in a user experience, and where and when it should appear" (p. 264).

A byproduct of the relationship between UX and content is UX writing. McConnell (2022) points out that "there's a lot of overlap between content design and UX writing (in fact, some organizations use the terms interchangeably)" (p. 2).

Torrey Podmajersky (2019) established a clear definition for UX writing as "the process of creating the words in user experiences (UX): the titles, buttons, labels, instructions, descriptions, notifications, warnings, and controls that people see." Tham, Howard, and Verhulsdonck (2023), for their part, describe UX writing as "a process involving writing, designing, thinking, iterating, strategizing, and developing content using different technologies to create a desirable user experience" (p. 8). Contributing to the elements that establish a difference between content design and UX writing, Carr (2021) explains that excellent UX writing is a component of the collaborative work that content designers are expected to perform at Meta.

While UX writing as a content creation process is often linked to content design, technical writing is not—a point we expand on in the section

"Placing the Entry-Level Technical Communicator as a Technical Content Professional." Like Collins (2021), we promote an understanding of content design as the disciplinary area that supports the development of both in-product content and technical content.

Just like content design has been adopted by tech companies because of the term's connection to UX design, content operations (particularly as ContentOps) is also advancing as an area of the discipline of content partially because of its perceived similarities with DevOps and other operational models in the industry's vernacular (e.g., DesignOps, ResearchOps, FinanceOps, AIOps).

Content Operations

For some content professionals, content operations is a term that includes *all* the operations of content (including content strategy and content design), while others use it to describe the operationalization of content, which takes place after content has been designed and/or a strategy has been developed. When Carlos was working with Rahel Bailie on the chapter "Operationalizing Content Creation" (Bailie & Evia, 2024), they regarded the relationship between content strategy and content operations with the following analogy that Rahel has been using for years in lectures and interviews (see Bailie, 2019):

- Content strategy is like the process of designing and creating the blueprints and plans for a new house. It is a process that can take from one to several months.
- After the content strategy is developed, a period of implementation follows. Following the comparison to building a house, this would be the actual construction process that normally takes several months and is longer than the strategy. (Note: this implementation period includes activities within the realms of content design and, as we discuss next, content engineering.)
- Content operations is the long process, of several years, that happens after the strategy has been implemented. It includes the day-to-day activities that enact the strategy. In the case of comparing it to building a house, it is the process of living in the house and performing routinary chores.

Rachel McConnell (2022) writes about the complementarity of content strategy and content operations. Acknowledging that it can be hard to tell them apart, she offers the following explanation for their relationship:

> (I)f content strategy is the engine, then ops is the oil, making sure that all the strategic elements work together. Strategy is a plan of action to achieve an outcome, and ops is the enabler of that plan.
>
> *(p. 13)*

Along the same lines, Colleen Jones (2019) argues that "well-executed content operations" should be combined with content vision and strategy to deliver "the right content for the right customer in the right touchpoint at the right time" (p. xiii). Jones provides the following definition for content operations:

> Content operations is the behind-the-scenes work of managing content activities as effectively and efficiently as possible. Today, content operations often require a mix of elements related to people, process, and technology.
>
> *(p. 162)*

Similarly, Scriptorium, for their part, defines content operations as "the engine that drives your content lifecycle" (p. 30).

Carrie Hane (2019), too, has defined the term as "the people, processes, and technologies that allow an organization to implement its content strategy to efficiently produce and effectively deliver content." She traces the origins of content operations to marketing teams and argues that "(l)ike content strategy itself, content operations has to rise out of the silo of one group within an organization to have maximum impact."

The book *Content Operations from Start to Scale: Perspectives from Industry Experts*, edited by Carlos (Evia, 2024), is built on the following definition of content operations provided by Rahel Bailie, which we find particularly useful and aligned with the way in which we define the disciplinary areas of content in this book:

> Content Operations (ContentOps) is the implementation of a strategy that incorporates people, processes, and technologies to optimize content production, so that organizations can leverage their content as business assets while retaining content quality.
>
> *(p. 4)*

The main purpose of the 2024 edited collection on ContentOps was to present perspectives from industry experts to inspire more advanced conversations and research projects related to this content-related area. The collection promotes an understanding of content operations as concerned primarily with governance and orchestration, an understanding that we promote throughout this book as a way to distinguish content operations from other disciplinary areas of the content discipline. With the exception of chapters authored by Carlos and Jason Swarts, all other contributions to that book came from industry experts and not full-time academics. That kind of industry–academia collaboration has also been present in content strategy and content design but is missing in the final area of the discipline of content that we will cover in this chapter: content engineering.

Content Engineering

Early usage of the term "content engineering" can be traced to the content consultancy Stilo Technology Ltd. (Content Engineering Stilo Technology Ltd., 2002) in the early 2000s as well as to academic publications from computer science researchers in Austria. In his exploration of the term, Siegfried Reich (2004) posited content engineering as an emerging discipline with the following abstract:

> There is a gap, one may call it a 'content crisis' even, between the growing opportunities for content creation and the increasing needs raised by content consumption. Content Engineering as a discipline may be a step forward to bridge that gap.
>
> *(p. 206)*

Reich proceeded to define content engineering as "the application of systematic and quantifiable approaches (concepts, methods, techniques, tools) along the content value chain, i.e., content acquisition, value adding, and distribution and delivery, in order to support re-use and tailorability for media-rich publications" (p. 208).

Although the main terms in Reich's definition—content reuse and tailorability (now better known as personalization)—and Stilo's description—XML, application programming interfaces (APIs), and other technologies—became common practices for practitioners working on the intersection of technical communication and the discipline of content, the term did not regain traction until the mid-2010s.

It is important to mention the computer science disciplinary area of document engineering, which has its own annual symposium—DocEng—under the Association for Computing Machinery (ACM). The DocEng ACM website defines it as follows:

> Document engineering is the computer science discipline that investigates systems for documents in any form and in all media. As with the relationship between software engineering and software, document engineering is concerned with principles, tools and processes that improve our ability to create, manage, and maintain documents.
>
> *(DocEng, 2024)*

Robert Glushko and Tim McGrath (2008) propose an approach to document engineering that "is mostly consistent" with the work of technical writers. They emphasize that their interpretation of the term includes documents that "are likely to be used more often by an application or web service than by a person" (p. 36).

Joe Gollner (2014), writing in the edited collection *The Language of Content Strategy*, revisited the concept of content engineering as "the application of engineering discipline to the design, acquisition, management, delivery, and use of content and the technologies deployed to support the full content lifecycle" (p. 72). Gollner even forecasted that "the discipline of content engineering represents the context within which content strategists, and indeed everyone involved in the content lifecycle, will operate from now on" (p. 73).

Cruce Saunders, alone and with his team from the content consulting agency Simple[A], has become the key proponent of content engineering as an area in the discipline of content—particularly as it intersects with technical communication. In the booklet *Content Engineering for a Multi-Channel World*, Saunders (2015) builds on Gollner's definition of the term, as he presents content engineering as "the discipline of organizing and shaping the structure and application of content, especially digital media, within technical environments" (p. 13).

Saunders also guest-edited an issue of *Intercom*, the magazine of the former Society for Technical Communication, on the topic of content engineering. In that issue, Ann Rockley (2019) listed the following tasks—which simultaneously seem connected to Reich's terminology in 2004 and forward-thinking for 2019 for a content engineer:

- Identifies how content varies based on customer needs and where each need arises in the customer journey
- Identifies how content can be modularized so that it can be automatically reused (mixed and matched) to meet customer needs
- Develops format-free, structured content models so that content can be written in a consistent way and automatically published to any channel (mobile, Web, print, wearables)
- Defines the structure of the content management system (CMS) repository so that it supports authoring and content retrieval
- Develops metadata to tag all content modules for personalized content retrieval
- Develops business rules to identify how content should be assembled automatically upon customer request
- Determines how to optimize the content creation process
- Determines how content has to be structured and tagged to support chatbots, natural language generation/automatic content generation, and AI (p. 9).

As of this writing, ZipRecruiter (2025) reports that the average salary for a content engineer is $116,615 per year in the United States. LinkedIn lists several job ads with the title of content engineer, but some of those have requirements and expectations that look more like software developers (or document engineers in the DocEng tradition) than content professionals.

There is a need to reconnect industry practices and academic interests in this area. Inspired by his close collaboration with preeminent content engineers working on the DITA Technical Committee (including Eliot Kimber and Robert Anderson), Carlos proposed to Rebekka a research project focused on exploring the academic profile of a content engineer. The project, which at that stage included a colleague from the Virginia Tech Department of Computer Science, became this book. During our work, we discovered that the profile of a content engineer is very different from that of the average technical communication undergraduate student in a US-based institution—with a background primarily in the humanities. We acknowledge that this part of our project is incomplete, but it took us in the direction of studying in which areas of the discipline of content, then, can a recent graduate in technical communication perform.

Placing the Entry-Level Technical Communicator as a Technical Content Professional

Regardless of the area under which leaders in academia and industry put the essential activities that are fused in the discipline of content, the majority of these activities take place at an advanced, strategic level in organizations and are not necessarily what an entry-level technical communicator would likely undertake. Our extensive exploration of books, whitepapers, blogs, webinars, and podcasts dedicated to defining and describing the activities necessary to develop and maintain a healthy life cycle for enterprise content, in addition to our conversations with leaders in the content industry, led us to one clear takeaway—that technical content creation, what we tend to think of as *content development*, was not considered its own disciplinary area and was often not accounted for as a key activity in other areas (a key reason being that content creation is governed by the principles and rules defined in other areas). While in-product content creation (frequently described as UX writing) is a key activity accounted for in content design, technical content creation is not.

So this left us to wonder where, then, did technical content creation—the primary activity of entry-level technical communicators working in organizations that have adopted workflows and technologies that treat content as computable data—belong in this fusion of content-related areas.

Also looking at where to place the task(s) of content development in the fusion of content-related areas, Collins (2021), as we mentioned before, suggested that content design is concerned with the creation and curation of content. Those who work in content design have job titles such as content marketers, UX writers, and technical writers. Setting extra boundaries around this area, McConnell (2022) provides a broad definition of content designers as "those who create product content."

These definitions of content design made sense to us—and locating technical content creation or development within this area gave us a way to talk not only about the entry-level technical communicator position but also about more advanced positions in other content-related areas—positions for which graduates can begin to prepare for in their education programs. Content design, then, might be best understood as the disciplinary area that supports the development of both technical content and in-product content. While some industry leaders may not agree with this description of content design, we find a clear distinction between the disciplinary areas, particularly content strategy and content design, useful for educators, students, and others wanting to learn about essential roles and activities at both entry and advanced levels within the content discipline.

Figure 2.1 articulates what we see as the relationship between the areas within the discipline of content. We offer this figure and the descriptions that follow as a way to operationalize our use of the terms throughout the book.

After a content strategy is developed, a period of implementation follows. This period includes iterative processes and activities in the areas of content engineering and content design that are orchestrated by content operations. We acknowledge that these descriptions are simplistic and that reducing whole disciplinary areas to a couple of sentences will exclude ideas. However, we think that these disciplinary areas might be further understood as follows:

- **Content strategy is about planning.** This disciplinary area articulates the plan and requirements for content as a valuable, computable asset. It focuses on understanding business and customer goals and, based on that understanding, planning for the creation, distribution, and governance of content.
- **Content engineering is about implementing.** This disciplinary area oversees the technical implementation of the content strategy plan and requirements. It applies engineering principles and approaches "to the design, acquisition, management, delivery, and use of content and the technologies deployed to support the full content lifecycle" (Gollner, 2014).
- **Content design is about producing.** This disciplinary area enacts the plan for creating in-product and technical content. Content designers working with technical content produce that content in accordance

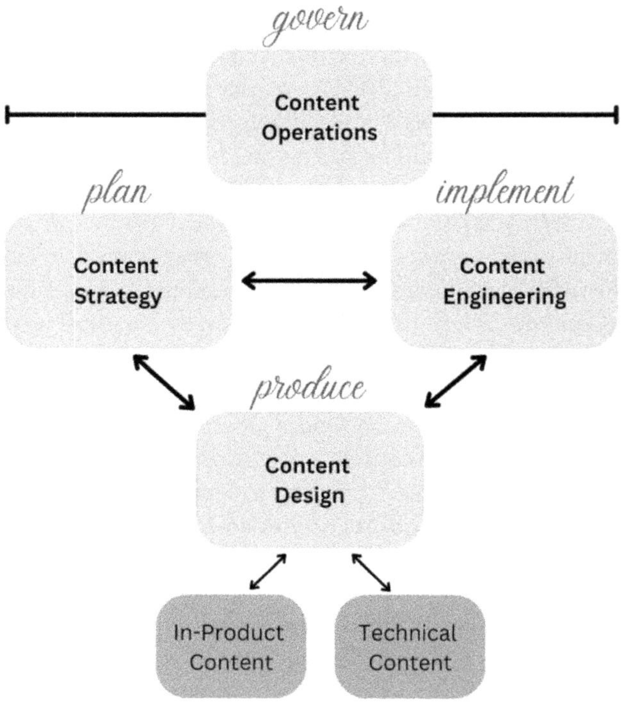

FIGURE 2.1 Disciplinary Areas within the Discipline of Content.

with principles, rules, and models defined in the content strategy and implemented via content engineers in content creation, management, and publishing tools.

- **Content operations is about governing.** This disciplinary area focuses on the people, processes, and technologies necessary for enacting the content strategy and achieving business goals; these activities include those connected to content engineering and content design.

As educators concerned with preparing students for new roles as content professionals, we find it useful to have this more structured way to understand roles and activities that support the life cycle of enterprise content. This approach gives us a shared way to talk about the skills and qualities graduates need for entry-level content creation positions (both technical and UX writing) and for moving into more advanced positions within the areas of content strategy, content operations, and content engineering. It gives us a way to think about the foundational skills and qualities that students need to be competitive for entry-level positions and the more advanced skills and qualities that our programs might be well positioned to cultivate given existing strengths of faculty and curricula. Ultimately, this combination of areas allows us to see and emphasize the importance of content-related work in digital transformation efforts and to investigate how technical communication programs can prepare students for these complex professional roles.

Digital Transformation Initiatives

The need to define the discipline of content and its related areas is a direct response to the need to bring strategic, scalable approaches to digital transformation initiatives. Writing about the McKinsey methodology for helping organizations work and succeed in an environment involving ever-evolving and emerging technologies, Eric Lamarre et al. (2023) define digital transformation, increasingly empowered by AI, as follows:

> A digital and AI transformation is the process of developing organizational and technology-based capabilities to allow a company to continuously improve its customer experience and lower its unit costs and over time sustain a competitive advantage.
>
> *(p. xiv)*

As GatherContent (2023), a content operations platform company, has said, "In the same way that we have standard frameworks for building software, or manufacturing cars—ContentOps calls for the same level of structure to be applied to the production, delivery and maintenance of content." We see the discipline of content as the essential standard framework that enables organizations to achieve the transformation goals outlined by Lamarre et al. and, in particular, to achieve digital content transformation goals.

Drawing on common understandings, the Center for Information-Development Management (CIDM), the Data Conversion Laboratory (DCL), and the RWS Group (2022) have defined digital transformation as "the process of using digital technologies to create or optimize business processes to improve company and customer experience and meet changing customer demand and market requirements." It is a process of transforming or re-imagining an organization, often tied to a corporate initiative, through the adoption of new or updating of existing technologies and tools. Similarly, Amy Williams, Chief Operating Officer at DCL, in an episode of the *Content Strategy Experts* podcast hosted by Alan Pringle (2022), described digital transformation as "using digital technologies to create or modify business processes and your customers' experiences." It is about meeting continuously changing business needs and market requirements, with a focus on the future. In 2021, according to Colleen Jones (2023), organizations spent more than $1.59 trillion on digital transformations alone.

Importantly, digital transformation efforts have significant implications for content. Today's digital technologies afford new, powerful ways of creating, managing, and delivering content at scale. Adopting new technologies to transform business processes and better meet customer needs means also adopting workflows and technologies that treat content as computable data. It means using technology to enrich customers' experiences with the company, the brand—and ensuring that customers can find the information that they need when they need it on the device they need it. Paraphrasing Steve Jobs, Jones (2023) has argued, "if your organization defines the problem incorrectly by omitting the content factor, the digital transformation will not be close to a solution." Digital transformation has been around for decades, but only in recent times has the rapid pace of technological innovation meant that digital transformation must happen more frequently. And that has meant that innovations in approaches to the discipline of content must happen more frequently, particularly as generative AI and large language models (LLMs) are adopted by content teams.

In a 2022 webinar hosted by the CIDM, DCL, and the RWS Group (2022), representatives from the three organizations reported the results of a survey of content professionals on digital transformation. The survey aimed to "understand how various industries address the adoption of [content] processes, technology, tools, and information architecture" as part of business-driven digital transformation. The following comments are representative of the different ways in which survey respondents described the purpose of digital content transformation efforts underway at their organizations:

- Finding better ways to deliver content to users—what they need, when they need it.
- Migrating people and processes from manual repetitive tasks to automated, collaborative scalable solutions.
- Modernizing and aligning the delivery of products and a company's customers and ensuring there is a consistent experience and messaging.
- Using the dynamic and multi-dimensional character of digital technologies to solve previously unsolvable (content) problems.

A key goal in digital transformation across represented organizations was to improve content experiences for customers and to adopt more innovative ways of delivering content out—of working with chatbots, voice-activated systems, and other emerging content distribution channels. Chip Gettinger, a host representing RWS, articulated the primary question that organizations are asking as follows: "How do we precisely provide the content our customers and employees are looking for, or don't even know that they should be looking for?" Digital transformations are happening across industries and most often involve multiple content-focused business units within organizations, including technical documentation, marketing, customer support, and training.

Atherton and Hane (2018), in their book *Designing Connected Content: Plan and Model Digital Products for Today and Tomorrow*, argue that all organizations undergoing a digital transformation must be concerned with how they connect with customers and thus must be concerned with how customers experience their content. "Being digital first," they argue, "means being content-first" (p. xv). There are numerous business needs and market requirements driving digital transformation. Below, we explain those most impacting content activities in organizations.

- **The need to publish or serve content in multiple outputs accessible from a plethora of devices.** Strategic attention on this need increased significantly starting in 2017. Based on an analysis of investment patterns among its clients, Simple[A] (2019), a consulting company specializing in content intelligent services, noted that 2017 was when they saw business leaders begin "sponsoring various initiatives aimed at enabling contextually-rich content experiences to intelligently flow to an ever-expanding landscape of devices and consumers" (p. 4). In 2020, we wrote about this need as one of the primary forces shaping content activities today. In the afterword that we wrote for the book *Teaching Content Management in Professional and Technical Communication*, "Beyond Management: Understanding the Many Forces that Shape Content Today," we talked about how "smart technologies and faster, better Wi-Fi are changing how and when people access content and what they expect from that content" (Evia & Andersen, 2020, p. 217).

  The plethora of devices from which people now access content—from smartphones to conversational interfaces to laptops—has meant that organizations must have a content presence across those devices if they are to remain competitive in the market. And that content must be designed (presentation agnostic, modular, and highly engineered) such that it can be delivered in an array of assemblages to many different outputs (e.g., PDF, HTML, native structured data, mobile apps) and accessed in specific assemblages responsive to specific inputs, contexts, and devices. Furthermore, advanced digital transformation approaches can even focus on providing content as a service, which can be integrated with apps and user tools via APIs instead of using an output formatted and published by a specific source organization (e.g., the fictional RobotVac company can serve "raw" content via APIs that clients, third-party developers, and users can then automatically include in their apps and interfaces, as opposed to having to focus on a website or print documentation developed, locked, and branded by RobotVac).

- **The need to provide seamless content experiences across a customer's journey.** Perhaps the most urgent business need and market requirement driving digital transformations is the need for organizations to provide customers seamless content experiences. Simple[A] (2018) described the need for focusing on customer journeys as follows:

> Today's thriving enterprises recognize the value of content and the role it plays in the customer experience. Consumers now expect highly relevant, timely, personalized content at all touchpoints of the customer journey, on any device or interface, across any and all channels of their choosing. Organizations that successfully deliver on these expectations reap the financial rewards and become leaders in their industries, while those that do not struggle to keep up with the rapid rate of change we all see around us.
>
> *(p. 9)*

As we wrote in the afterword for *Teaching Content Management in Professional and Technical Communication*, "customers expect to be able to access the content they need any time, from any device, in the exact format they need it (pdf, web, video)," and they expect, as well, "a consistent, unified message for the same type of content regardless of where or how it is accessed" so that they do not get confused or led to make the wrong choice (Evia & Andersen, 2020, p. 219). To meet this need, organizations must build the infrastructure to support consistency in content experiences across content touchpoints, from research and purchase decisions to installation and use needs to troubleshooting and maintenance questions. In thinking about a customer journey, for example, a customer interested in purchasing a RobotVac might first "touch" marketing content in an email accessed from their smartphone; they then might later open that email on their laptop and click to learn more, at which point they might be taken to the e-commerce site where they then purchase a particular RobotVac model; after setting up the RobotVac using the printed Quick Start guide that came with the product, the customer might run into an issue and decide to converse with the help chatbot available from the product's mobile application. These are all content touchpoints that represent a customer's journey with product content.

Better understandings of customer journeys and content needs from different touchpoints have given rise to the concept of "omnichannel," which focuses on how communication channels and supporting resources can be designed and orchestrated to cooperate, building a coherent, evolving, cross-channel experience (OmnichannelX, 2024). Omnichannel focuses on creating unified customer experiences. The goal, Hane (2019) explains, is to ensure that customers have a positive experience with an organization's content across all their communication channels: "Content runs through the entire experience from the way a customer service representative talks to someone to the way the website anticipates questions to confirmation emails to chatbots to Google's search results and many places in between."

Planning seamless content experiences across a customer's journey requires departments responsible for content to collaborate—or at least join forces in developing a unified content strategy. Simple[A] (2019) stresses this point in its whitepaper, "A Guide to the Content Services Organizations":

> To some degree, everyone is a member of the 'customer experience team' as sales and marketing content, product documentation, support, and technical assets are reviewed and rated by potential customers throughout the whole customer lifecycle. Content shared between marketing, sales, technical communications, knowledge management, learning, localization, and others contribute to unifying pre- and post-sale customer experiences as customers choose where to start and where to go next within the content sets.
>
> *(p. 9)*

Digital transformations focused on developing an infrastructure to support seamless content experiences across a customer's journey require extensive and strategic planning; intense, purpose-driven collaborations; and abundant resources and expertise. The discipline of content represents an assemblage of best, proven practices for developing this infrastructure.

- **The need to support continuous product development with continuous content delivery.** To stay competitive in the market and meet customer needs, more and more companies producing digital products have adopted a continuous product development process. This process focuses on building digital products in small increments, emphasizing flexibility and responsiveness to customer feedback and allowing products to be released faster. Miro (2023), a digital collaboration software business, describes the process as "iterative and customer-centric" and involving "constantly gathering user feedback, analyzing data, and making incremental improvements to the product throughout its lifecycle, with each product version being released as soon as it is ready." They identify several benefits of continuous product development, including rapid iteration and faster time-to-market, enhanced customer satisfaction and product-market fit, data-driven decision-making and performance optimization, and increased adaptability to changing market demands.

The adoption of continuous product development workflows has meant that content must also be continuously developed, with technical writers working in close collaboration with developers and product owners. The impact on content creators, Simple[A] (2019) suggests, is that they are being "asked to produce more than ever before, faster than ever before, all the while fundamentally changing publishing practices and workflows" (p. 4).

The most common workflow for continuously developing software is an Agile model of development, which enables software teams to "deliver work in small, but consumable, increments, rather than wagering everything on a 'big bang' launch"; an Agile process is founded on five key principles: iterative development, customer collaboration, adaptive planning, continuous feedback, and empowered cross-functional teams (simplilearn, 2024). Agile, which emphasizes flexibility and responsiveness to changing market conditions, has brought significant changes to publishing practices and content workflows. One of those changes is the adoption of docs-as-code approaches to content development and management in software environments. These approaches enable technical writers "to treat content components as computer code" and enable "a culture where writers and developers both feel ownership of documentation, and work together to make it as good as possible" (Holscher, 2024).

Many content developers, mainly those working with technical content, have embraced Agile practices that enable them to treat content components as computer code. Anne Gentle (2017) lists the following criteria for authors who are involved in projects treating docs as code:

- Store the doc source files in a version control system
- Build the doc artifacts automatically
- Ensure that a trusted set of reviewers meticulously reviews the docs
- Publish the artifacts without much human intervention.

- **The need to adopt a content ecosystem that allows for scaling up** (scalability) **and speeding up** (velocity)**.** As organizations grow through market expansion, acquisitions, and mergers, and as customers' requirements for content evolve, organizations need to be prepared to efficiently produce more content delivered to more channels. Building a content infrastructure that allows for scaling up in volume without slowing production for many organizations has thus become a top business priority, though one that is not easily realized. Sarah O'Keefe (2021) described the challenge this way:

> Company growth magnifies the challenges of information enablement. When you grow, you add products, product variants, markets, and languages—and each of those factors adds complexity. Process inefficiencies in your content lifecycle are multiplied for every new language or customer segment. As a result, content *scalability*—increasing content throughput without increasing resources—becomes critical.

The Scriptorium (2022) "Content Ops Manifesto" describes content scalability as an organization's ability to expand their "content lifecycle to process more content, more content types, more channels, more languages, and more variants" (p. 33). Key to this ability is a strategic approach to eliminating what Scriptorium calls "friction," which "refers to productivity impediments—processes that require human intervention" (p. 32). Examples of "friction" include repetitive tasks such as copying and pasting information and formatting information manually. The more these tasks are automated, the easier it is for organizations to scale up and produce more content fast. At the foundation of a content infrastructure that allows for scaling and speeding up is modular, structured content, which makes it easier not only to automate tasks but also to instantaneously publish more content and content variants in a range of outputs to a range of channels.

The numerous business needs and market requirements driving digital transformations and impacting content activities have brought on what some have called "a content revolution"—the need for a radically different way of thinking about how content—at scale—is planned, designed, and engineered. Simple[A] (2019), for example, has insisted on the need to elevate the nature and value of content in digital transformation processes. In a whitepaper articulating the value of content services teams in digital transformation, the organization argued, "We are standing in the midst of a significant revolution in the management of organizational knowledge and its expression as content"; "This revolution touches the publishing and engagement systems that handle content, the organizational structures and roles that produce and manage content, and every other part of the content ecosystem" (p. 3). Whereas the disciplinary areas of the discipline of content bring clarity to the activities necessary for carrying out this revolution, the key enabler of digital transformations across people, tools, and processes—and at the core of the discipline of content—is structured content.

Structured Content—Foundational for Digital Transformations

The business needs and market requirements driving digital transformations are not possible to meet without semantic structured content. Alan Pringle (2022) of Scriptorium underscored this point in a 2022 episode of *Content Strategy Experts* podcast, "Structured Content: The Foundation for Digital Transformation." He argued, "structured content is the building block, the basis, whatever you want to call it, for doing these digital transformation and innovation projects." Structured content, as a publishing workflow, "lets you define very consistently organized content in your documents, programmatically, so a human doesn't have to do it"; semantic tagging of that content then adds "a layer of intelligence" that facilitates both search and delivery.

Semantic structured content is modular content that is human and machine readable. Referring to it as "intelligent content," Rockley, Cooper, and Abel (2015) define it as content that is "designed to be modular, structured, reusable, format free, and semantically rich and, as a consequence, discoverable, reconfigurable, and adaptable" (this definition is based on an earlier definition by Rockley, 2008). It is content, according to Atherton and Hane (2018),

> that gets broken into small parts so that those parts can be recombined one way on a desktop web page, another way in a mobile app, and a third way when, say, heard through a voice interface like Amazon's Alexa.
> *(p. 12)*

It is the foundation of digital transformations—touching the needed components of people, processes, and tools—because of its versatility, portability, and manipulability.

These qualities are made possible through structure and semantic markup. Each content module (or content chunk) adheres to a specific structure—a specific set of software-enforced rules governing allowable elements, attributes, and relationships. These rules, defined in standards such as DITA, enforce consistent structure across similar pieces of information, such as a recipe or a product description. Each content module also includes semantic markup in the form of elements and attributes that describe the different kinds of content in the module. Semantic markup—or tags—are human and machine readable. The markup allows for machine processing of whole content modules (such as a task topic that gives instructions on how to replace the battery in a RobotVac) and the elements within them (such as an image example or troubleshooting tip). Metadata, a form of semantic markup, is data that describes the content chunk but is not a part of the content chunk; it is useful for directing content to one audience but not to another, to one type of device but not to another, and so forth. What each module does not include is markup that defines its presentation; rather, presentation is defined by stylesheets (e.g., XSLT, CSS) that assign presentation (or format) values to structural elements of the module based on where it appears, which may be in a mobile app, on a website, in a PDF, or even served via an API.

The affordances of structured content are vast. Modules can be seamlessly integrated and quickly assembled into large, complex documents; they can be simultaneously delivered to chatbots, portal sites, mobile apps, and other channels; they can be dynamically delivered as personalized content experiences; they enable continuous content development and publishing; and they allow organizations to automate repetitive tasks (reduce "friction"), scale up content development and publishing quickly, and provide high-quality source content to generative AI bots. Given its vast affordances, Atherton and Hane (2018) view the adoption of structured content as critical to an organization's future success:

> When structured around real-world concepts and connections, content scales up without breaking.... When you separate content and structure from presentation, you're in great shape to design interface representations for any new device that comes along. And when the content is described with metadata, right down to the last attribute, then every new device—even those without screens—can offer a completely tailored experience based on the same content. Chatbot? Sure. VoiceUI skill? Why not. Heck, people don't even have to come through your digital front door to get at your content. Do a Google search for movie show times in your area. You'll get the information you need without ever leaving the search results. That's not the future; that's today.
>
> *(p. 11)*

Similarly, Joe Gollner (2018) has suggested that as emerging channels such as chatbots and voice assistants take hold, structured content is going to require ever more semantic precision and portability. Such affordances do not come without extensive planning, design, and engineering—the strategic activities articulated in the maturing disciplinary areas of content strategy, content design, content engineering, and content operations.

For digital content to be usable, useful, and accessible in all the ways both users and stakeholders want, it must be modeled and engineered before it even exists. As Rebekka (Andersen, 2014) wrote in an article that examined rhetorical work in organizations that had adopted structured content, this strategic content framework "governs what content can be written and how it must be written, structured, and tagged to support reuse and dynamic publishing" (p. 120). That is, processes for creating, managing, and delivering content—including all the ways content assets relate or might relate—must be defined upfront, as part of the framework that governs the content ecosystem. Ann Rockley (2011) articulated this need for process and control as such: "Everything is planned, everything is identified, everything is controlled—this is how you get from beginning to end—from research and analysis all the way to users receiving content." The need for strategic planning at this level and scale has given rise to what we call the discipline of content.

For organizations that need to manage and produce enterprise-level amounts of content, building the strategic framework to support structured authoring workflows is a massive undertaking, and thus relatively few organizations have fully (and successfully) implemented a digital content transformation that accomplishes the goals described above. The CIDM, DCL, and RWS Group found in their 2022 survey of content professionals on digital transformations that only 7.6% of represented organizations had completed a full digital transformation; 2% of represented organizations were in the planning stage, 42% were in the implementing stage, and 23% were in the refining stage. Nearly half (47%) of these transformations, in their varied stages, were part of a corporate-wide initiative, 28% were part of a unit-specific initiative (e.g., technical publications, training, customer support, marketing), and 16% were part of a shared initiative with another business unit. Securing the resources required to support transformations—tools and technology, budget, time, staff, and training—was a major challenge across represented organizations. Without sufficient resources, the greatest challenges that content teams face, as reported in the 2022 Adobe Technical Communication Industry Survey findings report (Adobe Systems Incorporated, 2022), include difficulty "maintaining a single source of truth while creating, managing and delivering content," facilitating efficient review and collaboration workflows, converting existing content from legacy formats, implementing intelligent content reuse, and adopting new content formats such as chatbots, microcontent, and augmented reality.

Despite challenges around securing resources, the vast majority of survey respondents indicated that their business units had prepared for (or were preparing for) transformations in varied ways and in varied stages.

These ways included, among others, developing a content reuse strategy, planning for moving to structured content, developing a taxonomy and taxonomy management strategy (asking how teams were categorizing content, how people were finding content, how teams could make content more accessible), evaluating opportunities to reduce redundancy, analyzing content, and modeling content. These activities are essential to building the framework that ultimately governs the development, management, and delivery of content—they comprise the "plan of action" for a transformation, or what we and others identify as key components of the disciplinary area of content strategy.

It is clear through industry surveys and conversations among leaders that planning for digital content transformation is a must for organizations that produce products and services used by a wide customer base.

How AI Fits into the Content Professional's Toolkit

In the content industry, the launch of ChatGPT on November 30, 2022, and subsequent generative AI platforms since then has been met with both excitement about the seemingly endless possibilities for integrations with content operations and fear about loss of job security and prospects for technical writers. No doubt generative AI platforms are changing the activities and roles of technical communicators, as we have observed in the flurry of AI-focused webinars, podcasts, whitepapers, blogs, articles, and conference presentations hosted or published between 2022 and 2024. The webinar series *Content and Coffee* co-hosted by Scott Abel and Patrick Bosek has, for example, featured numerous leaders in the AI and content space discussing a range of topics from impacts of generative AI on technical communication work to strategies for creating effective prompts to use cases for generative AI in content workflows. Prominent content leaders such as Rahel Bailie, Sarah O'Keefe, Colleen Jones, Casey Jordan, Jason Kaufman, Anthony Medeiros, Jack Molisani, Ellis Pratt, and Cruce Saunders have all weighed in on current and future uses of AI in the discipline of content. Launched in October 2023, the *Content + AI* podcast hosted by Larry Swanson—a podcast in which Larry has conversations with experts on both AI and content practice—has also kept technical communicators abreast of AI trends.

While organizations are indeed integrating AI into business, development, content, and other workflows, and while many organizations are seeing significant savings and efficiency gains, it is too early to say in any definitive sense how AI will ultimately impact the activities and roles of technical communicators and content specialists. Most of what leaders have had to say to date has been speculative or based on specific use cases. Studies of generative

AI in technical communication work contexts broadly and in technical content teams specifically are certainly underway, but as of fall 2024 published results were not yet available, and cumulative studies that point to definitive trends, use cases, and impacts were even further off.

In July 2024 and January 2025, we did see the first special issues of journals focused on AI and technical communication, which included a handful of articles relevant to, though not specific to, technical content work contexts (see Carradini, 2024, 2025). But just as we did when paradigm-shifting trends such as single sourcing, structured authoring, and content management became widespread in the profession—before game-changing books such as Ann Rockley and Charles Cooper's *Managing Enterprise Content: A Unified Content Strategy* (2012) or Rahel Bailie and Noz Urbina's *Content Strategy: Connecting the Dots Between Business, Brand, and Benefits* (2013) helped codify best practices and before scholarship on trends, practices, and implications for research and teaching emerged—we can look to leaders of content solutions consultancies, leaders in the content solutions vendor space, and innovative leaders of content teams for early indicators of how AI is impacting technical communication work and what we can expect in the near future.

There is no question that AI will be integrated into all aspects of content workflows, sooner rather than later for organizations with mature content operations. Larry Swanson, in his inaugural episode of the podcast *Content + AI* (2023a, October 23), suggested to listeners that "AI's going to be able to help us across the span of content practice" and that "No matter what kind of content work you do, you'll soon be using AI in any number of ways." In his third episode (2023b, November 8), guest Noz Urbina, Co-Founder of Urbina Consulting & OmnichannelX and self-identified Human-centric Prompt Engineer, talked extensively about how his consulting agency is helping UX teams leverage AI throughout the research stage of the design process, and he urged listeners to be prepared for AI to be part of every stage of the content life cycle. Chief Executive Officer of Scriptorium Sarah O'Keefe, featured guest of Swanson's fourth episode (2023c, November 12), compared generative AI to spellcheckers:

> It would not occur to anybody in this day to write content without using a spellchecker. It's just part of the groundwater. It's part of the fundamental tool set, that bag of tools that you're sitting around with when you actually go to create content. So, in many ways AI is going to be that spellchecker.

At the time we wrote this chapter, these were all predictions by well-known leaders in the content space. It was a time of experimenting and working through what AI could do and how it could do it.

Integrating AI into content workflows has not come without its challenges. AI chatbots trained on public data sets, such as OpenAI's ChatGPT, Google's Gemini, Microsoft's Copilot, and Anthropic's Claude, have presented a particular challenge for content teams. Because these chatbots can generate biased or inaccurate responses, organizations cannot risk publishing technical content that has not been carefully vetted and tested by humans. As Tham, Howard, and Verhulsdonck (2022) have argued, AI chatbots "learn from the databases on which they're trained, and if those databases contain inaccurate, biased, libelous, or even scandalous material, then the bots will reproduce that content unless corrected and fact-checked by humans" (p. 211). Further, because such AI chatbots do not have access to new product content, as it has yet to be written, and because they are trained on limited data available freely on the internet, they are prone to generating inaccurate instructions that could cause all kinds of harm or damage, from a hard drive overheating to a TV crashing to the floor to a medical device being used incorrectly. These are significant concerns for technical writers and their organizations because of their legal, safety, usability, and UX implications.

One workaround that technical writers have adopted to avoid the risk of using chatbots trained on public data sets is to feed chatbots their own content and ask them to work only with that content. Chatbots can be trained on that content and produce, given effective prompting, reliable outputs void of bias and inaccuracies. Identifying trends in a data set, synthesizing notes, generating summaries of reports, and helping to revise and edit drafts are all tasks that public chatbots can do remarkably well when working with content that a writer has fed them.

As of the fall of 2024, most content teams did not appear to be using public chatbots in content workflows. Rather, most teams were using modified public LLMs or using custom-built LLMs within their organizations. In a November 2023 webinar on "Using ChatGPT and Gen AI in TechComm" (Abel, 2023), Ellis Pratt spoke about how some organizations were using public LLMs that they had modified by adding new data (importing content) from their organization and then changing the weightings of different parameters. While the chatbot is still connected to the public LLM to learn grammar, voice, tone, and other language skills, it can be restricted to providing answers from only the organization's content. The second approach he mentioned, which we expect most large organizations to adopt, is to build chat interfaces and custom language models (or hire a company to do this for them). Doing so allows for training the language model on a specific corpus of content and for walling off that data and content. The customized chatbot is then fed only vetted source content and outputs responses based on that content.

With more advanced-level technical communicators now having the experience of assessing digital transformation needs and developing a content strategy and operations plan, we are likely to see an increase in strategic positions aligned with content strategy, content engineering, and content operations in years to come.

In academic conversations related to technical communication and the discipline of content, this topic has also taken over the titles of presentations and panels at conferences. However, as of this writing, there are not many scholarly research articles published about the impact of generative AI on technical communication. Carol Reeves and JJ Sylvia (2024) have published a review of research in this specific area of our field. We expect many more research projects exploring generative AI from technical communication scholars in the near future.

In this chapter, we intentionally did not dive into the weeds of defining AI or explaining how AI works. There are several resources available that explain key concepts of AI, including LLMs, natural language processing, neural networks, deep learning, and machine learning. For resources directly connected to AI and writing, we recommend the special issues of the *Journal of Business and Technical Communication* focused on AI tools in technical communication (see Carradini, 2024, 2025) as well as Larry Swanson's *Content + AI* podcast. Ask any generative AI platform to explain AI and how it works, and you will receive an informative, detailed response that is as good as most available published descriptions. In Chapter 3, we highlight what leaders in the content and AI space have identified as important trends impacting the activities and roles of technical communicators working in modern organizations.

References

Abel, S. (Host). (2023, November 7). Using ChatGPT and gen AI in TechComm: Ellis Pratt discusses research survey results, insights, and practical applications [Webinar]. In *The Content Wrangler Webinar Series*. Heretto and The Content Wrangler. Retrieved from https://www.brighttalk.com/webcast/9273/597514.

Adobe Systems Incorporated (2022). *Adobe technical communication industry survey 2022: Current and future state – Key findings* [White paper]. Retrieved June 27, 2024, from https://www.adobe.com/products/technicalcommunicationsuite/whitepaper-technical-communication-survey-findings-2022.html.

Andersen, R. (2014). Rhetorical work in the age of content management: Implications for the field of technical communication. *Journal of Business and Technical Communication*, *28*(2), 115–157.

Atherton, M., & Hane, C. (2018). *Designing connected content: Plan and model digital products for today and tomorrow*. San Francisco, CA: New Riders.

Bailie, R. (2019). Content strategy. *Technical Communication*, *66*(2), 121–124.

Bailie, R. (2024). Defining content operations. In C. Evia (Ed.), *Content operations from start to scale: Perspectives from industry experts* (pp. 13–24). Blacksburg: Virginia Tech Publishing.

Bailie, R., & Evia, C. (2024). Operationalizing content creation. In C. Evia (Ed.), *Content operations from start to scale: Perspectives from industry experts* (pp. 43–74). Blacksburg: Virginia Tech Publishing.

Bailie, R., & Urbina, N. (2013). *Content strategy: Connecting the dots between business, brand, and benefits*. Laguna Hills, CA: XML Press.

Barker, D. (2007, November 24). The four disciplines of content management. Retrieved July 1, 2024, from https://deanebarker.net/tech/blog/four-disciplines-of-content-management/.

Batova, T., & Andersen, R. (Eds.). (2016). Content strategy—A unifying vision [Special issue]. *IEEE Transactions on Professional Communication*, *59*(1).

Bridgeford, T. (2020). *Teaching content management in technical and professional communication*. New York: Routledge.

Carr, E. (2020, October 6). Why we're moving from content strategy to content design. Retrieved June 26, 2024, from https://design.facebook.com/stories/why-were-moving-from-content-strategy-to-content-design/.

Carr, E. (2021, December 6). Writing the metaverse: Meta's VP of content design on the future of the discipline. Retrieved June 26, 2024, from https://design.facebook.com/stories/writing-the-metaverse/.

Carradini, S. (Ed.). (2024). Effects of artificial intelligence tools in technical communication pedagogy, practice, and research, Part 1 [Special issue]. *Journal of Business and Technical Communication*, *38*(3).

Carradini, S. (Ed.). (2025). Effects of artificial intelligence tools in technical communication pedagogy, practice, and research, Part 2 [Special issue]. *Journal of Business and Technical Communication*, *39*(1).

Casey, M. (2023). *The content strategy toolkit: Methods, guidelines, and templates for getting content right* (2nd ed.). Berkeley, CA: New Riders.

Center for Information-Development Management (CIDM), the Data Conversion Laboratory (DCL), and the RWS Group (2022, August 25). *Digital transformation survey 2022: Results, analysis, and projections* [Webinar]. Retrieved June 19, 2024, from https://www.infomanagementcenter.com/product/digital-transformation-survey-2022/.

Collins, J. (2021, June 17). The maturing content discipline [Blog post]. Retrieved June 26, 2024, from https://www.linkedin.com/pulse/maturing-content-discipline-john-collins/.

Content Engineer Salary. *ZipRecruiter*. Retrieved January 17, 2025, from https://www.ziprecruiter.com/Salaries/Content-Engineer-Salary.

Content Engineering Stilo Technology Ltd. (Management). (2002, December). *Database and Network Journal*, *32*(6), 3+. Retrieved January 21, 2025, from https://link.gale.com/apps/doc/A95449305/AONE?u=unilanc&sid=bookmark-AONE&xid=31ca6e72.

Crestodina, A. (2022). *Content chemistry: The illustrated handbook for content marketing*. Chicago, IL: Orbit Media Studios

Crow, M. (2002). A new American university: The new gold standard. Retrieved July 1, 2024, from https://president.asu.edu/read/a-new-american-university-the-new-gold-standard.

D'Agostino, F. (2012). Disciplinarity and the Growth of Knowledge. *Social Epistemology*, *26*(3–4), 331–350. https://doi.org/10.1080/02691728.2012.727192.

Deacon, R. (2002). Truth, power and pedagogy: Michel Foucault on the rise of the disciplines. *Educational Philosophy and Theory*, *34*(4), 435–458. https://doi.org/10.1111/j.1469-5812.2002.tb00518.x.

DocEng (2024). ACM symposium on document engineering. Retrieved June 2, 2024, from https://doceng.org/.

Evia, C. (2024). Operationalizing content creation. *Content operations from start to scale: Perspectives from industry experts*. Blacksburg: Virginia Tech Publishing.

Evia, C., & Andersen, R. (2020). Afterword: Beyond management: Understanding the many forces that shape content today. In T. Bridgeford (Ed.), *Teaching content management in technical and professional communication* (pp. 213–231). New York: Routledge.

Feldman, S. (2020, April 11). A list of content disciplines. Retrieved April 15, 2024, from http://www.sarafeldman.com/effective-content/a-list-of-content-disciplines/.

GatherContent (2023). What is 'ContentOps' and why should you invest in it? Retrieved June 26, 2024, from https://gathercontent.com/content-operations.

Gentle, A. (2017). *Docs like code: Write, review, test, merge, build, deploy, repeat*. Austin, TX: Just Write Click.

Getto, G., Flanagan, S., & Labriola, J. (Eds.). (2023). The people, practices, and technologies central to content strategy [Special issue]. *Journal of Technical Writing and Communication, 53*(4).

Getto, G., Labriola, J., & Ruszkiewicz, S. (2019). *Content strategy in technical communication*. New York: Routledge.

Getto, G., Labriola, J., & Ruszkiewicz, S. (2023). *Content strategy: A how-to guide*. New York: Routledge.

Glushko, R. J., & McGrath, T. (2008). *Document engineering: Analyzing and designing documents for business informatics and web services*. Cambridge: The MIT Press.

Gollner, J. (2014). Content engineering. In S. Abel & R. Bailie (Eds.), *The language of content strategy* (pp. 72–73). Laguna Hills, CA: XML Press.

Gollner, J. (2018, April 25). The marriage of structure and semantics: DITA and master content models. Paper presented at the CIDM CMS/DITA North America Conference, Denver, CO.

Hane, C. (2019, July 25). Content operations—The missing link between strategy and delivery [Blog post]. *Zen—A Tanzen Blog*. Retrieved June 17, 2022, from https://www.tanzenconsulting.com/blog/2019/7/24/content-operations-the-missing-link-between-strategy-and-delivery.

Halvorson, K. (2008, December 16). The discipline of content strategy. Retrieved July 20, 2022, from https://www.alistapart.com/articles/thedisciplineofcontentstrategy.

Halvorson, K. (2018). Ten years later: The discipline of content strategy. Retrieved June 26, 2024 from https://www.braintraffic.com/blog/ten-years-later-the-discipline-of-content-strategy.

Halvorson, K. (Host). (2023, September 26). Keri Maijala, LinkedIn - Building and leading a content practice at LinkedIn (No. 60) [Audio podcast episode]. In *The Content Strategy Podcast*. Brain Traffic. Retrieved from https://www.braintraffic.com/podcast/episode-60-keri-maijala-linkedin-building-and-leading-a-content-practice-at-linkedin.

Halvorson, K., & Rach, M. (2012). *Content strategy for the web* (2nd ed.). Berkeley, CA: New Riders.

Holscher, E. (2024). Docs as code. Retrieved June 26, 2024, from www.writethedocs.org/guide/docs-as-code/.
Jones, C. (2019). *The content advantage (Clout 2.0)*. San Francisco, CA: New Riders.
Jones, C. (2023, September 4). 4 characteristics of modern digital transformation—and content implications [Blog post]. *The content science review*. Retrieved from https://review.content-science.com/4-characteristics-of-modern-digital-transformation-and-content-implications/.
Jorgensen, E. (2023). *Strategic content design: Tools and research techniques for better UX*. New York: Rosenfeld Media.
Keshtcher, Y. (2022, May 31). UX writing hub: Salary survey report. Retrieved June 27, 2024, from https://uxwritinghub.com/ux-writing-salary-survey.
Kissane, E. (2011). *The elements of content strategy*. New York: A Book Apart.
Lamarre, E., Smaje, K., & Zemmel, R. (2023). *Rewired: The McKinsey guide to outcompeting in the age of digital and AI*. Hoboken, NJ: Wiley.
Long, M. (2020, November 26). UX writing hub: The rise (and role) of content design. Retrieved June 27, 2024, from https://uxwritinghub.com/the-rise-of-content-design/.
Lovinger, R. (2007, March 27). Content strategy: The philosophy of data. *Boxes and Arrows*. Retrieved July 1, 2024, from https://boxesandarrows.com/content-strategy-the-philosophy-of-data/.
McConnell, R. (2022). *Leading content design*. New York: A Book Apart.
McGrane, K. (2012). *Content strategy for mobile*. New York: A Book Apart.
Miro. (2023). What is continuous product design? Retrieved June 26, 2024, from https://miro.com/product-development/continuous-product-design/.
O'Keefe, S. (2021). Content scalability: Removing friction from your content lifecycle. Retrieved July 3, 2024, from https://www.scriptorium.com/2021/05/content-scalability-removing-friction-from-your-content-lifecycle/.
Omnichannel X (2024). What is omnichannel? Retrieved June 26, 2024, from https://www.omnichannelx.digital/what-is-omnichannel/.
Podmajersky, T. (2019). *Strategic writing for UX*. Sebastopol, CA: O'Reilly.
Pringle, A. (Host). (2022, August 15). Structured content: The foundation for digital transformation with guest Amy Williams (No. 125) [Audio podcast episode]. In *The Content Strategy Experts*. Scriptorium. Retrieved from https://www.scriptorium.com/2022/08/structured-content-the-foundation-for-digital-transformation-podcast/.
Pullman, G., & Gu, B. (2008). *Content management: Bridging the gap between theory and practice*. Amityville, NY: Baywood Publishing.
Reeves, C., & Sylvia, J. J. (2024). Generative AI in technical communication: A review of research from 2023 to 2024. *Journal of Technical Writing and Communication*, *54*(4), 439–462. https://doi.org/10.1177/00472816241260043.
Reich, S. (2004). Content engineering: Bridging the gap between content creation and consumption. In U. K. Wiil (Ed.), *Metainformatics. MIS 2004. Lecture Notes in Computer Science* (pp. 206–211), vol. 3511. Berlin: Springer.
Rigolot, C. (2020). Transdisciplinarity as a discipline and a way of being: complementarities and creative tensions. *Humanities & Social Sciences Communications*, 7(100), 1–5.

Rockley, A. (2003). *Managing enterprise content: A unified content strategy*. Indianapolis, IN: New Riders Press.
Rockley, A. (2008). What is intelligent content? The Rockley Group. Retrieved June 26, 2024, from https://www.rockley.com/articles/What%20is%20Intelligent%20Content.pdf.
Rockley, A. (2011, February 16). Intelligent content eye-opener: Content strategy. Paper presented at the Intelligent Content Conference 2011, Palm Springs, CA.
Rockley, A. (2019). Content engineer roles and responsibilities. *Intercom*, 9–11.
Rockley, A., & Cooper, C. (2012). Managing enterprise content: A unified content strategy (2nd ed.). Indianapolis, IN: New Riders.
Rockley, A., Cooper, C., & Abel, S. (2015). *Intelligent content: A primer*. Laguna Hills, CA: XML Press.
Rose, R. (2023). *Content marketing strategy: Harness the power of your brand's voice*. London: Kogan Page.
Saunders, C. (2015). *Content engineering for a multi-channel world: Enabling the next generation of customer experiences*. Austin, TX: Simple[A].
Savage, G. (2023). Profession. In Y. Han & J. Buehl (Eds.), *Keywords in technical and professional communication* (pp. 213–219). WAC Clearinghouse: University Press of Colorado.
Scriptorium. (2022), *Content transformation: An introduction to enterprise content ops and content strategy*. Research Triangle Park, NC: Scriptorium Publishing Services, Inc. Retrieved June 26, 2024, from https://www.scriptorium.com/content-transformation/.
Simple[A]. (2018). A new content order for the multi-channel, multi-modal world. *Intercom*, 9–11.
Simple[A]. (2019). *A guide to the content services Organization* [White paper]. Retrieved June 26, 2024, from https://simplea.com/Publications/Whitepapers/A-Guide-to-the-CSO.
Simple[A]. (2016, August 16). What is content engineering? Retrieved July 16, 2025 from https://simplea.com/resources/articles/what-is-content-engineering.
Simplilearn. (2024). What is agile? Understanding Agile methodology and principles in software development. Retrieved June 26, 2024, from https://www.simplilearn.com/tutorials/agile-scrum-tutorial/what-is-agile.
Swanson, L. (Host). (2023a, October 23). Content + AI introduction – episode 0 (No. 0) [Audio podcast episode]. In *Content + AI*. Larry Swanson and Elless Media. Retrieved from https://contentandai.com/intro/.
Swanson, L. (Host). (2023b, November 8). Noz Urbina: The RAUS Method for accelerating content projects with AI – episode 3 (No. 3) [Audio podcast episode]. In *Content + AI*. Larry Swanson and Elless Media. Retrieved from https://contentandai.com/noz-urbina/.
Swanson, L. (Host). (2023c, November 12). Sarah O'Keefe: AI in technical communication and content strategy – episode 4 (No. 4) [Audio podcast episode]. In *Content + AI*. Larry Swanson and Elless Media. Retrieved from https://contentandai.com/sarah-okeefe/.
Tham, J., Howard, T., & Verhulsdonck, G. (2022). Extending design thinking, content strategy, and artificial intelligence into technical communication and user experience design programs: Further pedagogical implications. *Journal of Technical Writing and Communication*, *52*(4), 428–459.

Tham, J., Howard, T., & Verhulsdonck, G. (2023). *UX Writing: Designing User-Centered Content* (1st ed.). New York: Routledge. https://doi.org/10.4324/9781003274414
Verhulsdonck, G., Howard, T., & Tham, J. (2021). Investigating the impact of design thinking, content strategy, and artificial intelligence: A "Streams" approach for technical communication and user experience. *Journal of Technical Writing and Communication*, *51*(4), 468–492.
Wachter-Boettcher, S. (2012). *Content everywhere: Strategy and structure for future-ready content*. New York: Rosenfeld Media.
Walker, C. J. (2023, June 24). A brief history of content strategy. Retrieved July 1, 2024, from https://firehead.net/2013/06/a-brief-history-of-content-strategy/.
Winters, S. (2017). *Content Design*. London: Content Design London.

SECTION II

Roles, Activities, and Career Trajectories

3

CHANGING TECHNICAL COMMUNICATOR ROLES AND ACTIVITIES

A Quest to Understand Entry-Level Technical Content Roles and Activities

When we set out to learn how roles and activities are changing for technical communicators in organizations that are undergoing or have undergone digital content transformations, we were particularly interested in changes affecting entry-level technical communicators. We wanted to know how work contexts, needs, and expectations were changing for the kinds of positions for which graduates of technical communication programs were or could be applying.

As we have presented earlier in this book, our exploration of this topic prompted us to organize two workshops at industry and academic professional conferences. We also conducted 31 semi-structured interviews with 17 industry experts in one or more areas related to content and with 14 academic experts in programs that continually update curricula to align with industry practices. In addition to interviewing industry and academic experts, we talked to 12 recent graduates from programs offering a BA or BS in technical communication and one alumnus from a graduate MS program. A final data collection leg of our study focused on a survey aimed at managers and recruiters of technical communication professionals. The survey asked a series of questions focused on the following:

- hiring needs and challenges
- skills and qualities managers and recruiters look for in new hires and why
- skills and qualities necessary for advancement to more strategic positions

DOI: 10.4324/9781003451730-6

- suggestions for how undergraduate programs might evolve to better prepare students for content-focused jobs

Fifty-two self-identified managers or recruiters who met our criteria for inclusion completed the survey.

After analyzing responses from the managers who had completed our survey and the managers, consultants, and recruiters who had participated in our interviews, we realized that the roles and activities described were primarily those of technical content creators—or technical writers. The entry-level position had indeed changed, but it was still largely a technical writing position. When describing the activities of this entry-level position, respondents did not talk about modeling content, developing taxonomies, designing user experiences, or other strategic activities that are often discussed in the context of other areas fused in the discipline of content. Those activities characterized more advanced positions, such as content strategist or taxonomist, that entry-level technical writers often move into with more experience, training, and expertise.

As we describe in Chapter 2, locating technical content creation or development within the disciplinary area of content design enables us to accomplish the following:

- Find a way to talk about the entry-level technical communicator position in organizations that have adopted digital content transformation workflows
- Develop language to describe more strategic activities and roles associated with advanced positions in other areas of the discipline of content
- Have a shared way, for educators, to describe the foundational skills and qualities that graduates need in order to be competitive for entry-level positions within content design (both technical writing and UX writing) and for eventually moving into more advanced positions within content strategy, content operations, and content engineering.

Importantly, this entry-level position remains the main door into the discipline of content. This position represents how graduates of technical communication programs get in the arena and, ideally, gain meaningful agency in and exert influence on their organizations over time. Thus, for educators, understanding the contexts shaping the entry-level position and what organizations need and expect from entry-level technical communicators is paramount.

In this chapter, we report what we learned through our interviews with industry leaders and through our survey of hiring managers about the ways in which entry-level technical communicator roles and activities are changing in organizations that are undergoing or have undergone digital content transformations. These leaders included managers, consultants, and recruiters. While most managers who participated in our research worked in the area of content design (managing teams of writers), some worked in content strategy, content operations, and content engineering, straddling different roles. All of the consultants with whom we spoke worked in the areas of content strategy, content operations, and content engineering. It was clear through our conversations with leaders that the majority of content-focused entry-level positions in their organizations—or in the organizations for which they had completed work as consultants—were in the area of content design, with a majority of positions titled "technical writer" or "user assistance specialist."

The 17 industry leaders whom we interviewed had a combined 472 years of experience in the field, most of those years in leadership roles. We identified these individuals as experts based on their reputation in the field and on the extent to which they were considered thought leaders—whether that be through shaping best practices in the discipline of content or being widely cited and called upon for content-focused workshops, webinars, and conference talks. Twelve of those experts owned a content services or solutions company, carrying titles such as Founder and CEO, President, and Founder and Principal. The other five served in roles such as Director of User Assistance, Senior Director of Information Architecture, and Senior Manager of Content Strategy—all employed by Fortune 100 or 500 companies. Figure 3.1 shows the number of years our interviewees had worked in the field; many of them had started out as writers, engineers, or developers and had moved into leadership roles as they built specialized expertise in—and contributed to development of best practices in—the discipline of content. Notably, 82% of our interviewees had worked in the field for 25 years or more.

All 52 of our survey respondents were in management roles, with close to 50% of their position titles being some flavor of Technical Documentation Manager and another 17% being some flavor of User Assistance Manager. While 4% had worked in management for five or fewer years, 38% had worked in management for 25 or more years. Not surprisingly, given concerns over the aging workforce that we report on in Chapter 4, only 8% of our survey respondents had worked in the field for ten years or less. Nearly half had worked in the field for 21 or more years (the impending wave of retirees, suggests some, points to a potential hiring crisis in the industry in the near future). Figure 3.2 presents a breakdown of the number of years that respondents, who were all in management roles, had worked in the field.

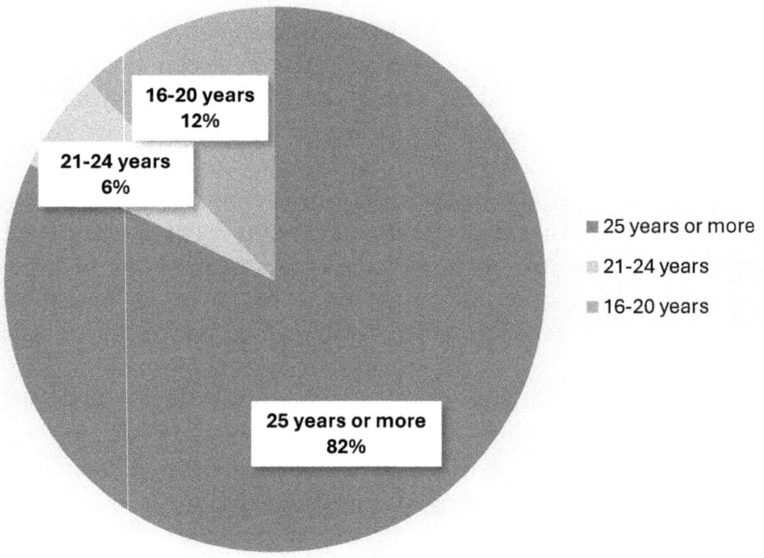

FIGURE 3.1 Interview Participants—Breakdown of Years Industry Experts Had Worked in the Field of Technical Communication.

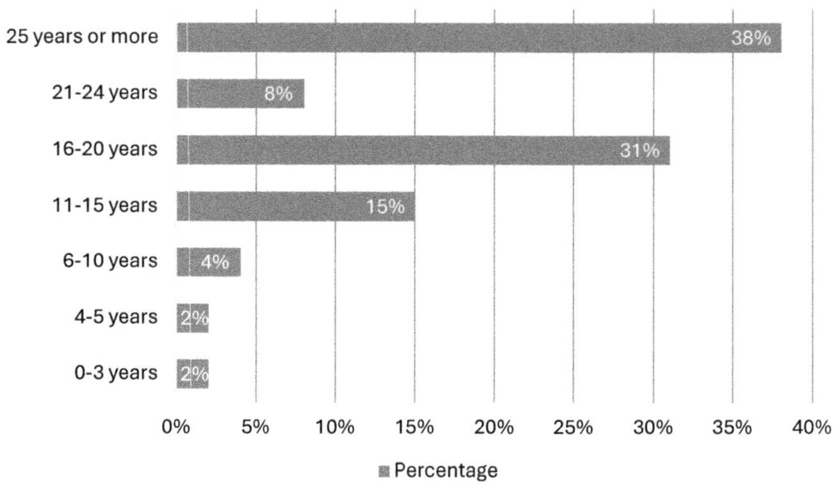

FIGURE 3.2 Survey Respondents—Breakdown of Years Managers Had Worked in the Field of Technical Communication.

In addition to talking with industry leaders, we talked with 14 administrators or faculty in undergraduate programs in technical communication that continually update curricula to align with industry practices. We identified these individuals as experts based on their reputation in the field, record of publishing research focused on industry practices, and commitment to developing curricula responsive to industry changes. Ten interviewees were serving in administrative positions at the time, heading undergraduate programs in technical communication, and four interviewees were faculty members.

Our main goal in asking questions about roles and activities was to better understand the work contexts, needs, and expectations for the kinds of positions for which graduates of technical communication programs might apply. To gain the perspective of new professionals who had recently completed undergraduate degrees in technical communication (and one having just completed an MA degree), we also interviewed 13 program alumni. Of the 13, five were working in entry-level technical writing positions for organizations that had adopted or were in the process of adopting digital content transformation workflows. In this chapter, we share profiles of each of the five alumni, with each profile capturing what the alumni had to say about their goals upon graduation, their current roles and responsibilities, the publishing environments of their organizations, and the skills and qualities needed to move into more advanced content-focused positions. We present each profile in a breakout box in the *Technical Content Activities and Impacts on Roles* section.

Table 3.1 presents the questions related to changing activities and roles that we asked of these different groups. Although we did not ask questions about specific areas within the discipline of content, and we did not use "discipline of content" as a term, we did recruit participants who worked primarily with technical content to gather perspectives on how roles and activities are changing in the discipline of content.

The themes summarized in this chapter are specific to roles and activities in the technical content space, which has been both a leader in driving digital content transformations and an example of how content treated as a business asset can optimize business processes and meet changing customer demand and market requirements. As a reminder, throughout this book we use the term *technical content* to refer to any type of content that is treated as computable data for the purpose of being both human and machine readable, adaptable, reusable, and portable. Given the scope of our work at the intersection of the disciplines of content and technical communication, we did not ask about changing roles and activities in the other disciplines that also intersect with content, such as marketing or computer science.

TABLE 3.1 Survey and Interview Questions Focused on Changing Roles and Activities

Interviews—Industry Leaders	• Can you talk a bit about what it means to be a technical communicator today? • How do you see technical communicator roles and activities changing? At the entry level? At the advanced level?
Interviews—Academic Leaders	• From your perspective, what does it mean to be a technical communicator today? • How do you see technical communicator roles and activities changing? At the entry level? At the advanced level?
Survey—Managers	We asked the following closed-ended questions. • Does your team primarily work with technical content (e.g., online help, user guides, API documentation, release notes, UI text)? • Which descriptions best represent the technical content activities in your organization? Check all that apply. We asked the following open-ended questions. • In the organizations in which you have worked in the past ten years, how have entry-level technical communication roles and activities changed? What have been some of the driving factors behind these changes (e.g., processes, technologies, mergers?) • In the organizations in which you have worked in the past ten years, how have advanced-level technical communicator roles and activities changed? What have been some of the driving factors behind these changes (e.g., processes, technologies, mergers?)
Interviews—Alumni	• What are your roles and responsibilities as [position title]? What kinds of activities do you engage in as a [position title]? • How would you describe the publishing environment of your organization or your particular team or unit? How is content authored, managed, published/delivered, and with what tools?

Being a Technical Communicator Today

When we asked industry and academic leaders to talk about what it means to be a technical communicator today, we heard descriptions that echoed what we might have heard had we asked the question 30 years ago. They talked about the many ways in which a technical communicator is (or needs to be) a bridge between subject matter experts (SMEs) and end users, a bridge between the product that someone is trying to use and their ability to use it, an advocate for the user, and a strong writer and communicator. These descriptions cut across industries and specializations, and they appear in varying forms across widely referenced definitions of technical communicator.

In communicating about technical or specialized topics and communicating how to do something, the technical communicator serves as a bridge between experts and end users and as an advocate for the user's ability to successfully achieve their goal, whatever that might be. They make technical or specialized information usable and accessible. Cleary (2021) captures these themes well in her working definition of technical communication, based on her review of definitions offered by the field's professional organizations:

> [t]echnical communication involves communicating, explaining, and helping people to learn, understand, use, and interact with specialized concepts or technologies. Technical communicators design interventions to help people to use technology, services, or information, safely and effectively. These interventions take many forms, from print content to knowledge bases, developer documentation, e-learning courses, video, and myriad digital formats. This is both a humanistic and a technical discipline because of its focus on enabling people to interact with devices, services, and content.
>
> *(p. 8)*

Cleary's definition is useful in that it conveys both the explanatory and instructional aspects of the kinds of communication deliverables that we create, and it emphasizes our goal of helping people—of enabling people to understand and use devices, services, and content to achieve their goals. In her definition, we are both a bridge and an advocate.

Several leaders who participated in our study described the role of the technical communicator as that of an enabler. It is, as one prominent consultant[1] put it, someone who "creates the information or creates the bridge between the complex product or service that you're trying to use and the ability to actually use that thing." The goal is to enable people to use products and services successfully. Another frequently cited consultant made a useful distinction between what technical communicators do and what other kinds of communicators do. She suggested that technical communicators create enabling material, whereas other kinds of communicators might create marketing material or sales material—what she referred to as persuasive content. Enabling content, for this consultant, is "all the informational, technical non-persuasive material that enables your product to be sold or your service to be provided or your application to deliver the messages." She added that it can be explanatory or instructional with the goal of facilitating action of some kind.

Creating successful enabling content requires technical communicators to advocate for the user experience, to understand what users need, when, how, and why. This was a key point that came up across our conversations with leaders in industry and academia. In the context of organizations that have undergone digital content transformations, though, being an advocate for the user requires being engaged in system-level content design decisions—often as a content strategist or content engineer. While what it means to be a technical communicator has not changed all that much, the processes, methods, and digital technologies that we use to design content experiences have. And as Rebekka has argued previously (Andersen, 2014), so too have the nature and location of rhetorical work. In more traditional document development contexts, agency to decide what to write, how to write it, and how to format it often lies with the writer. In content ecosystems that rely on modular components, those who have primary rhetorical agency—those positioned to design user experiences with devices, services, and content—are those who work in the areas that we identify as content strategy and content engineering. They design the system that governs how content is written, structured, and tagged for reuse and multichannel publishing. Those in these strategy-level roles are thus best positioned to be what several leaders called "champions" for the user.

Being a champion for the user in modular content production environments means moving beyond just being an advocate to being someone who fights for the user at the highest level of decision-making. One consultant described this role as follows:

> And so for me, when you are the person who has to communicate how something works or how someone can apply something and get value out of it, you often run into all the issues that that person is going to run into and you add so much more value, if you can really become, even go beyond what I would say is advocacy for that user but actually become a champion for that user. So you need to be the advocate first because you need to understand them without becoming them. But then you need to be the person who will go and fight for that person to really be able to achieve the success that they're trying to achieve or minimally to achieve the success that the product originally promised them. And fighting means being a content leader.
>
> —*CEO/Founder of a communications and content solutions consultancy*

The role of champion requires technical communicators to have agency within the content system, or what Kimball (2016) referred to as the "content machine." This role is not an entry-level one or one that focuses on content creation. It is, rather, a person who has the skill set, knowledge, and drive to move from advocate at the writer level to champion at the system level. This is the professional profile of someone who works outside the core team and influences and drives initiatives in areas such as engineering, product management, and design. Another consultant also described the role of the "champion" in response to our question about what it means to be a technical communicator today. He suggested that the notion of champion, someone who engineers customer experiences, "takes a high-level view of the systems through which content flows" and advocates for the user not at the level of documenting a thing but at the level of the systems that govern the creation, management, and delivery of content.

So, while what it means to be a technical communicator today—a bridge, an advocate, and a strong writer and communicator—has not changed all that much, how these goals are operationalized in digital transformations has, quite radically. Entry-level technical communicators can and do advocate for users within the limited context of communicating with SMEs, learning about the products they are documenting, and writing and managing structured content. As technical communicators move into more advanced positions in organizations that have adopted digital content transformation workflows, advocacy takes on the form of champion—being a content leader who drives content strategy and may oversee content operations. It is at this level that technical communicators have the most agency over the interaction between the product and the user and the product itself—through engaging with the process by which the product gets designed and developed. It is also at this level that technical communicators are well-positioned to define and drive initiatives and activities with social justice goals, gaining buy-in from management and other leaders and assessing impact through data-driven methods. We are seeing more and more job titles such as Vice President of Content Strategy or Director of Content Operations as organizations recognize the need for digital transformation content leaders.

Working in these environments, technical communicators—whether in advanced or entry-level roles—must be system thinkers and collaborators and must be responsive to change and ready to adapt quickly. Baehr (2019) suggests that technical communicators must be "agile communicators" if they are to succeed in today's rapidly changing, technologically complex communication environments, particularly those supported by Agile production processes. "Communicators who are agile," Baehr argues, are able to

- "assess situations as they change and adapt quickly and effectively"
- "target incoming changes, employ their research and knowledge management skills to understand it and its effects, and then adjust and flex their communication skills to encompass or work with it"
- "use the technological tools they need to support high-performance" (p. 9).

For Baehr, this is what it means to be a technical communicator today. As rhetorical agency moves from the lone writer to the interdisciplinary team charged with designing and overseeing content ecosystems, technical communicators must be prepared to work and think at a system level that engages stakeholders from across an organization and to apply technical expertise to both understand complex products and systems and contribute meaningfully to content ecosystems.

Here, we want to emphasize the importance of technical expertise. Because technical communicators now produce many different kinds of products supported by increasingly complex, technical processes and tools, they need more technical knowledge and know-how than was once required. They need this expertise to work effectively with engineers and developers, particularly in Agile development environments; to understand principles and workflows of software and hardware development; and to learn and effectively use new tools quickly. The following comments are representative of the kinds of points made by many of the leaders with whom we spoke:

> So to be a technical communicator involves a number of things. The least of which is lack of fear or fearlessness. I should say of technology, right? So being able to break it down without fear of breaking it and being curious and fearless in your approach to understanding how things work and what information people need to know in order to be successful with their products and services.
>
> —*Principal consultant and co-founder of an authoring solutions company*

> There's a lot more of the technical side of things, of just being comfortable with XML structured authoring of various kinds, whether that's DITA or anything else. Also not being afraid of that technical aspect, which I think, is another thing that I see—a lot of people go into writing because they think they won't have the technical aspects, and then they

see a bunch of DITA code or something in front of them and they get a little intimidated.

—*President and owner of a content solutions consultancy*

I don't think many roles today will allow you to not be technically savvy, so even though you may not be the developer and you may not be writing the code that goes into the software product, you have to know enough technically about the product that's being created in order to be able to talk to the developers. I think frankly the way we're delivering information means you have to know enough about the backend of how information is transmitted, collected. So I'm thinking about things like structured authoring or the fact that you might need to know enough about building websites so that you can be the person who posts the information that customers are going to need. Maybe you're not dealing with the SQL server and whatever else, but you know enough HTML and CSS to at least present information.

—*Professor and director of corporate relations at a university's department of technical communication*

I'm one of two technical writers in an entire organization that probably has around 1000 employees working in various divisions, and so I would guess that there are maybe eight to 10 teams that push stuff out every year that involve me. The teams are varied in what they focus on, so to be a technical communicator today, you have to navigate all these complex technical domains and try to create documentation without having a strong knowledge of the technology.

—*Senior technical writer*

What we learned through our many conversations with industry and academic leaders is that being a technical communicator today means being a bridge, advocate, and strong writer and communicator—characteristics that have long defined the role. But it also means being a champion for the user experience, an agile communicator, a technical specialist, and an interdisciplinarian. These characteristics speak to changes in the processes, methods, and technologies that we use to design content experiences, as well as to changes in the nature and location of rhetorical work—that is, "the process by which we create and shape our communication for specific audiences in specific contexts for specific purposes" (Andersen, 2014, p. 118). To advocate for users today, one must be able to think and work at a system level, collaborating and problem-solving with many interdisciplinary stakeholders.

Technical Content Activities and Impacts on Roles

A majority of the managers who participated in our research (100% of interviewees who were managers and 98% of survey respondents) led teams that worked primarily with structured, technical content in XML or Markdown environments. Outputs varied widely, but common deliverables included online help, user guides, application programming interface (API) documentation, release notes, and UI text. Although we did not ask survey respondents to indicate which industry they represented, responses suggested that most respondents worked in the technology, information technology services and solutions, and manufacturing industries, with documentation of hardware and software being a key job responsibility.

When asked to describe the technical content activities of their organizations, 77% of survey respondents indicated that their teams created primarily structured content, and 67% indicated that their teams published content in a variety of formats (e.g., dynamic delivery, print, electronic PDF, HTML, embedded help, video). The most common way teams managed content was through a source control system (e.g., SVN, Git), with content component management systems (CCMSs) and content management systems (CMSs) being the second and third most common, respectively. Figure 3.3 identifies the technical content activities of the organizations of managers who responded to the survey.

What these findings tell us is that many teams that have adopted structured content workflows are not just creating structured "text" content but also videos and/or animations as well as UI content. Slightly more than half of respondents, as well, indicated that they share content across multiple teams—a key affordance of structured content. Our findings align with the digital transformations survey findings of the CIDM, DCL, and

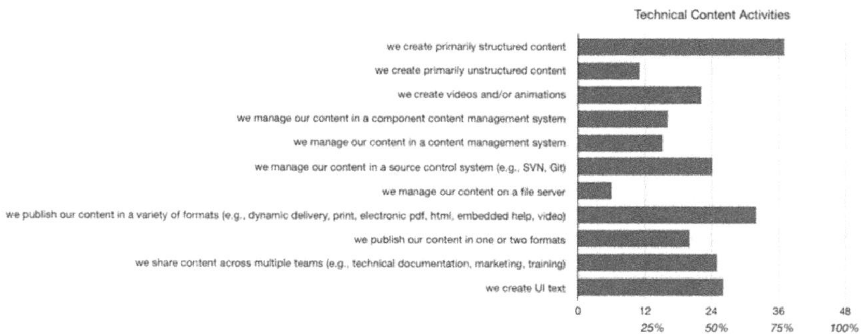

FIGURE 3.3 Technical Content Activities Reported by Survey Respondents.

the RWS Group (2022). The primary ways respondents of their survey managed content were through a CCMS, CMS, source control system, or file system on a network. In reporting results, Chip Gettinger of DCL highlighted the use of integrated systems, particularly CCMSs that integrate with digital asset management systems; this allows for digital assets such as videos to integrate with text assets such as structured content. Other recent surveys with which these findings align include the 2019 Technical Communication Benchmarking Survey of 600+ professional communicators, conducted by The Content Wrangler (Abel, 2018a, December 13), and the 2018 Census of Technical Communicators conducted by Saul Carliner and Yuan Chen (2018) and supported by the Society for Technical Communication (STC).

To identify themes in responses to our interview and survey questions about how technical communicator roles and activities are changing at the entry and advanced levels in organizations that have undergone digital content transformations, we used the qualitative analysis platform NVivo. In NVivo, we gathered and organized all responses to related questions (asked slightly differently in interviews than on the survey) and then read and thematically coded responses. Some interviewees talked about a particular activity or role multiple times during the interview; we coded activities and roles only once if they came up multiple times so that we could determine total unique cases of discussed activities and roles, which also allowed us to compare interview data with survey data.

The dominant themes that emerged in response to our questions about changing roles and activities were *structured authoring* and *Agile development*—interestingly, in his survey of 203 STC members, Lanier (2018) also identified these themes as dominant trends and issues impacting approaches to writing and designing content. Below, we report what our participants had to say about these themes. Connected to these themes are descriptions of the diverse products and content outputs that organizations now produce, and the software development processes as well as the tools and platforms that support content activities. Although we did not ask questions about the impact of generative AI and large language models on technical communicator roles and activities, and only a few research participants mentioned AI, we would be remiss not to address what we are now learning about the impact that AI is having on technical communicator roles and activities. We thus end this section with a synthesis of what we have learned through our review of AI-focused webinars, podcasts, whitepapers, blogs, books, articles, and conference presentations hosted or published between 2022 and 2024.

Structured Authoring

Definitions, Approaches, and Advantages

Structured authoring is a method for writing structured content, which we see as a foundational component of digital content transformations. In the field of technical communication, structured authoring is frequently achieved through the use of XML and, particularly, the standard known as the Darwin Information Typing Architecture (DITA). We have been working with DITA for a couple of decades, and Carlos was a member of the non-profit technical committee maintaining the DITA standard for several years. Therefore, discussion about structured authoring in this book leans heavily toward XML and DITA. However, many companies of diverse sizes have, with considerable limitations—particularly when it comes to managing multiple languages in a content repository or word/phrase-level filtering—implemented structured authoring workflows with JSON or even implemented lightweight markup languages like Markdown, reStructuredText, or AsciiDoc as their standard for managing and delivering structured content. Regardless of the standard, structured authoring is a modular approach to writing content that can be reused in multiple contexts and published in different formats. O'Keefe and Pringle (2017), in their widely referenced whitepaper, "Structured authoring and XML," define structured authoring as "a publishing workflow that lets you define and enforce consistent organization of information in documents, whether printed or online" (p. 2).

Most entry-level technical writers working in organizations that rely on modular components contribute to structured authoring workflows. Rather than create whole documents as writers once did in traditional publishing environments, writers create and review modular content in collaboration with other writers using tools that include editing apps or build systems, CCMSs or source control systems, and publishing processors. In some environments, writers work with an all-inclusive system such as Heretto CCMS, which allows teams to author, review, manage, and publish modular content in a variety of outputs and formats without having to toggle between tools. In other environments, writers work with multiple integrated tools, such as an XML editor like Oxygen, a CMS like Adobe Experience Manager, and a publishing processor, often one built on the DITA Open Toolkit. Hackos explained the DITA Open Toolkit, and its relationship with the DITA standard, as follows:

> The DITA Open Toolkit (DITA-OT) is an implementation of the OASIS DITA Technical Committee's specification for DITA DTDs and schemas. The Toolkit transforms DITA XML source content into deliverable formats such as PDF, HTML, and help systems. Although the

> DITA-OT is closely associated with the DITA specification, it is not a standard approved by the OASIS DITA Technical Committee and therefore should be considered as only one option for transforming DITA content.
>
> *(2011, p. 325)*

DITA was released as a set of design principles for creating "information-typed" modules at a topic level and for using that content in delivery modes such as online help and product support portals on the web (Day et al., 2001). A topic, as described by Bellamy, Carey, and Schlotfeldt (2011), is "a self-contained unit of information" that "covers only one subject" and "is long enough to make sense on its own, but short enough to stick to one point without expanding into other subjects" (p. 8). Don Day, one of the original developers of DITA as an XML experiment for managing documentation at IBM, explained that, when naming the standard, DITA "represented a great deal of messaging in a compact and memorable acronym":

- Darwin: for specialization and how things could "evolve" from a base
- Information typing: for representation of knowledge as typed units
- Architecture: a statement that this was not just a monolithic design but an extensible tool that could support many uses (Day, quoted in DITAWriter, 2016).

Iantosca (2025) has written about the ideas at IBM that generated DITA, which the company eventually donated as an open standard. DITA is currently maintained by the non-profit consortium OASIS, and "has evolved substantially since that initial donation to encompass a very wide scope of requirements indeed" (Kimber, 2012, p. 6). At the OASIS DITA Technical Committee, the standard continually evolves with the purpose "to define and maintain the Darwin Information Typing Architecture (DITA) and to promote the use of the architecture for creating standard information types and domain-specific markup vocabularies" (OASIS Darwin Information Typing Architecture TC, n.d.). Hackos summarized the key benefits of DITA for technical communicators as follows:

- A fully tested document type definition (DTD) or schema for XML-based authoring
- A community of developers investing in improvements to the DITA model
- An open-source toolkit you can use to produce your own output in multiple
- Media without having to invest in proprietary tools

- A thoroughly developed approach to information development originating with OASIS and now encompassing many other companies, large and small, that find value in a standards-based approach (2011, p. 9).

Furthermore, DITA's most cited business advantage is promoting the reuse of information quickly and easily across multiple deliverables, but other oft cited advantages include reducing the cost of maintaining and updating information, enabling continuous publishing, sharing information across the global enterprise, reducing costs of localization, and reducing technical debt caused by inadequate, incorrect, and unusable legacy information (Hackos, 2011, p. 10).

How information is organized within a DITA topic or module is predetermined by an external file that can be coded in the formats of XML Schema (XSD), DTD, or Regular Language for XML Next Generation (RELAX NG); that external file governs allowable elements, attributes, and relationships. The XSD, DTD, or RELAX NG file is referenced in each topic or module that determines whether that topic or module is valid—that is, whether it conforms to the rule set defined in the XSD, DTD, or RELAX NG file. An advantage of using DITA instead of custom-made implementations of XML is that those validating files are planned, developed, tested, and revised by the DITA Technical Committee of volunteer experts and released as part of the open standard. Once validated against the governing file, the content topic or module can be rendered into an output for end users. Importantly, structured authoring separates content from format; only at the stage of publishing is format applied to content via stylesheets (CSS or XSLT files, or a combination of those). This separation allows for content to be easily reused and updated, and it allows for multichannel publishing.

Figures 3.4 and 3.5 show different views of a basic DITA topic authored in the integrated development environment Oxygen XML Editor—a commercial product from the Romanian company SyncRO Soft based on the DITA-OT. This topic is an introductory content component to the fictional RobotVac cleaning system.

The DITA technical specification establishes that a topic can only have one instance of a short description, defined as an element that "represents the purpose or theme of the topic" and "is also intended to be used as a link preview and for search results" (OASIS, 2018). The DITA specification dictates that the "short description should be a single, concise paragraph containing one or two sentences of no more than 50 words." The sample RobotVac introductory topic from figures 3.4 and 3.5 has two paragraphs inside its short description. This is not valid according to the DITA

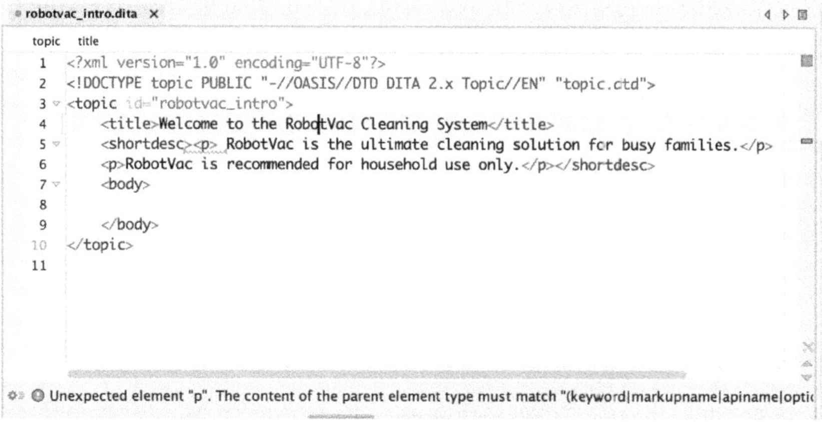

FIGURE 3.4 Sample Introductory Topic for the Fictional RobotVac System Created in Oxygen XML Editor. This View, Labeled as "Text" in Oxygen, Allows the User to Write and Edit the XML Code Directly. This Sample Topic Shows an Error Message When Validated against the DITA Definition Files.

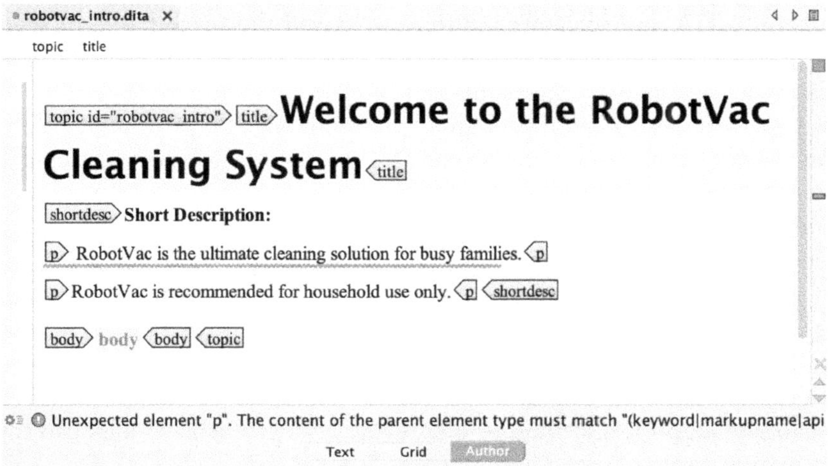

FIGURE 3.5 Sample Introductory Topic for the Fictional RobotVac System Created in Oxygen XML Editor. This View, Labeled as "Author" in Oxygen, Offers a "What You See Is What You Get"-Like View That Hides the XML Code. This Sample Topic Shows an Error Message When Validated against the DITA Definition Files.

definition files or schemas, and as a result, the DITA validation is giving the author an error message.

The technical specification published by OASIS (2018) is the authoritative reference for in-depth information about the DITA standard. For those interested in DITA who do not want to read the whole specification, we recommend JoAnn Hackos's (2011) *Introduction to DITA* (still relevant although a bit dated in its examples) and Carlos's (Evia, 2018) *Creating Intelligent Content with Lightweight DITA*. We include additional recommendations in Chapter 7 and in this book's companion GitHub repository.[2]

Widespread Adoption and the New Content Professional

In its annual survey of over 2,000 self-identifying technical communicators, Adobe (2018) found that adoption of structured content in industry had reached the halfway point, growing from 20% to 50% between 2012 and 2017. Similarly, in the 2018 issue of *Intercom* that shared results of the state of the technical communication industry census, Abel (2018b) reported that the creation of structured content was a growing trend on teams around the globe, with 51% of teams surveyed reporting use of XML to create content (55% of those teams reported using the DITA XML schema) (p. 37). In that same issue, Carliner and Chen (2018) reported on the work practices that most affect technical communicators: standards such as DITA and structured authoring played the biggest role in work projects (p. 15). Adoption has steadily grown since then, though no recent survey has examined the percentage.

Our interview and survey findings certainly align with those of Adobe, Abel, and Carliner and Chen. Moving from a document-based writing process to structured authoring was reported by hiring managers, consultants, and recruiters as the biggest change in technical communicator activities in recent years. At the same time, structured authoring is not a new method of writing, as many technical communicators have been creating, managing, and delivering technical content in this way for decades.

What is newer is the spread of structured authoring across content domains within an organization (across technical content, marketing, customer support, and more)—a strategic effort to achieve digital transformation goals of providing seamless content experiences across a customer's journey and publishing content in multiple outputs accessible from a plethora of devices. The need to create voice content and make it accessible via conversational technologies is also driving the need for more machine-processable and human-readable content. So too are customer demands for personalized content and videos, and company-specific implementations of generative AI platforms like ChatGPT. Structured content enables personalization and can be output in video format. In fact, in their 2022 survey of self-identifying technical communicators, Adobe (2022) found that a majority of respondents were working toward the goal of making content interactive and engaging by creating video tutorials, and they were doing this by merging technical content with video formats.

The widespread adoption of structured authoring to support digital content transformation goals has created the need for the discipline of content. As one consultant who owns a content solutions agency told us,

> ...we're at a renaissance period now where the necessities of true content intelligence are just starting to (make) sense inside of organizations. And organizations are starting to reorganize operations and their content supply chains to support this new way of working.

In addition to enabling a rethinking of the nature and affordances of content within an organization, structured authoring tools and technologies have enabled improved processes and workflows for content stakeholders. One hiring manager noted that structured authoring in a CCMS had significantly improved process efficiencies, "enforcing voice, style, grammar, and graphics consistency and introducing better work, review, and publish workflows." Another manager emphasized the improved efficiency and quality of reviews through the use of automated tools.

For entry-level technical communicators, the adoption of structured authoring and structured authoring tools and technologies has meant that they need to be ready to start creating content and contributing to content workflows soon after their date of hire. Numerous hiring managers spoke to this point in their survey responses, two stating the following:

> One of the big changes...has been that the entry-level people I currently work with immediately took training on the content management system and began working in it. The technology I currently work with is so complex that it is taking a very long time for the entry-level people to become proficient in it. It is somewhat challenging to find appropriate training tasks for them. Also, even the entry-level people have needed to be quite self-directed and self-managing, probably because there are fewer managers than in the past.
>
> —*Survey respondent*

> Earlier, entry-level technical communicators helped the seniors with their work. Most of the time, they were like shadows. They were expected to have language and written communication skills. Technical writing concepts were taught at work (with mentors guiding them through live

> projects). Now, entry-level technical communicators start working on doc projects. They are expected to be self-starters and work independently with minimal supervision.
>
> —*Survey respondent*

A common theme among managerial responses to our question about changing activities was that of reduced training resources and onboarding time. New hires in technical communication need to be ready to contribute, and that means that they need to come to the position with an understanding of the structured authoring paradigm. They need to understand writing as a process divorced from documents, and they need to think in terms of information typing, chunks, and structure. To succeed in a structured authoring environment, according to one consultant with whom we talked, writers need to "understand content as units or genres as small units of information that can be assembled in many ways to become larger genres"; they need to understand "reuse and multiple use of the same content in different structures." Jack Molisani echoed this point in a September 2023 episode of *The Content Strategy Experts* podcast. In response to host Bill Swallow's (2023, September 11) question about how roles are changing in content development, Molisani, president of a content professionals employment agency, commented:

> Used to be you'd see a job opening and it would say technical writer needed, must have FrameMaker and RoboHelp. Whereas now, it's like, all right, yes, we want someone who does structured authoring. We don't care which tool. Because really, once you've got the concept of structured authoring down it doesn't matter what content management system or structured authoring development tool you're using. Because if you know one, you can pick up the others.

BOX 3.1 ALUMNI PROFILE: KAT—TECHNICAL SPECIALIST FOR CONTENT SOLUTIONS CONSULTANCY

In 2018, Kat graduated with a BA in English from a large public research institution. She started her undergraduate journey as a Green Engineering major, completing enough courses to receive a minor in Green Engineering before changing her major to Professional and Technical Writing. During her senior year, she took courses on user documentation using the DITA standard and web development, continued to take engineering courses (including courses in basic programming), and completed a research assistantship focused on the implications of content management for technical communication education. These experiences spurred her interest in exploring career options as a technical content specialist.

With her engineering background and interest in writing documentation, Kat's goal upon graduation was to apply for positions that allowed her to work at the intersection of "the communication side of things" and technology and "the development side of stuff."

When she saw an internship position for a technical specialist at a content solutions consultancy, she immediately applied and was offered the position. She worked as an intern for six months before being hired full time. At the time of her interview, Kat had worked for her company for 14 months (including her time as an intern).

Kat described her roles and responsibilities as "shifting all the time" given the nature of the work that a consultant agency does. She started out working on marketing for the company to learn about the company culture and how things work. She was then assigned a more technical role that allowed her to work on DITA conversions and "do the actual technical writing and content strategy side of things." At the time of the interview, she had moved into what she called "more of a developer" role, focusing less on the day-to-day activities of the company and more on the technical side of DITA projects. She described her role as "helping other people develop strategies on how to write their content"—it was a role that she felt she was easily able to assume given her fluency in writing code and writing DITA content when she was originally hired as an intern. But she also emphasized the on-the-job need to "learn how to teach [developer skills] to yourself," something she had to do most days, such as when she would be tasked with writing a script.

Her organization specializes in helping clients develop and implement a content strategy, including building content models, converting legacy content to structured content, implementing a CMS, and coding and developing back-end XSLT style sheets. The consultancy does its own marketing as well, via social media and the web, whitepapers, and frequent webinars and conference presentations as well as podcast episodes. Kat explained,

> We have content to attract clients that we put out, but then we also have content that we author for our clients and it's a little meta: it's like we're writing technical documentation about their technical documentation—this is the way that you need to use this stuff to publish your technical content.

Kat described her team as "practicing what we preach"; they all knew DITA and wrote all company content in DITA (except for the company Wordpress blog) for publishing in multiple outputs. In commenting on the leadership structure of the organization, Kat explained that for each client project, a team with a manager, developer, and one or more technical specialists is put together, with the manager being the top of the "chain of command." The manager focuses on strategic vision and planning whereas the developer "tests and runs and fixes."

When asked about advancement pathways in her organization, Kat talked about how "working your way up in tech comm is all about experience." She described skills such as writing XSLT scripts, creating DTDs, and writing DITA content as skills that are difficult to teach and learn but skills that are essential to landing entry-level jobs in tech comm and moving into more advanced positions. For her, personally, she was positioning herself to advance by taking on projects that she knew how to do and welcoming opportunities to contribute to projects that were outside of her wheelhouse. She said, "when I come across something I don't know how to do, I go to [the lead developer], and everyday I learn a little bit more." She talked about approaching the content strategy side of things in the same way, welcoming opportunities to assess a client's needs and write a report articulating problems and recommended solutions. She would then seek feedback from the CEO and COO and use that to continue to build her skill set.

Those entry-level technical communicators who are ready to contribute to structured authoring workflows also need to understand the concepts of minimalism and classification. In minimalist writing, each unit of content has a purpose and answers one question, and it discards information irrelevant to that purpose, that question. In technical communication, the concept of minimalism is based on the work of John M. Carroll, who proposed an approach to "minimizing the obtrusiveness to the learner of training material—hence the term *minimalist*" (1990, p. 7, his emphasis). The three key aspects of Carroll's minimalist approach to instruction are the following:

1 Allowing learners to start immediately on meaningfully realistic tasks
2 Reducing the amount of reading and other passive activity in training
3 Helping to make errors and error recovery less traumatic and more pedagogically productive (p. 7).

In the 1990s, Carroll's work on minimalist documentation influenced the IBM team that developed the original DITA schemas for structuring content, and to this date, the DITA Technical Committee promotes the standard as an implementation of those principles for minimalist instruction.

Classification focuses on what to call units of content and how to categorize those units. Multiple managers and consultants with whom we talked emphasized the need for new writers to know how to classify and tag content so that people can easily find the information that they need when they need it. One consultant referred to this need as a library science mentality—"having a good understanding of how to categorize...information and how to help people find that information, I think, is becoming a critical aspect to the job." A senior director of user assistance described the need for classification systems and taxonomies from the perspective of a user looking for information:

> People are searching for the information they need, so findability is key. I have to be able to find what I need when I want it. You can give me a lot of material, I'm okay with that when I search for it, but I really just want that little piece at that moment. Based on how I search, the right content should appear, and that right content will depend a lot on how that content is tagged. The world is moving into machine learning, into machine translation, into content automatically being identified by machines as a concept or as a task or as a reference.

Familiarity with semantic markup, including metadata, and taxonomies is key to knowing how to classify and tag information for findability. This activity, as well, is becoming more common as the need to efficiently feed search engine optimization, artificial intelligence, and machine learning engines becomes more prevalent.

Developing expertise in the various components of structured authoring, from the separation of content from form to information typing to semantic markup and taxonomies, requires a fundamentally different way of thinking about writing, organizing, and designing documents and other deliverables. It requires, as Swarts (2020) argues, a "conceptual understanding" that "consists of understanding text structures, modular content, and reuse" and understanding "rhetorical characteristics of DITA-structured writing" (p. 156). Preparing entry-level technical communicators to contribute to structured authoring workflows, then, requires both a conceptual and applied approach that few have articulated in terms of a training curriculum. Swarts (2020) offers perhaps the most robust articulation, describing an approach to teaching structured authoring that focuses on structure, bilocational meaning through reuse, and translocation meaning through modeling, mapping, and metadata.

Whereas entry-level technical communicators in more traditional desktop publishing environments often assume one primary role, such as writer or editor, those working in structured authoring environments can have many roles. Because of continuous product development and the need to collaborate with product teams and other content teams, communicators can expect to be more involved in project management and deliverable management and prepare to participate early and throughout the product development process.

BOX 3.2 ALUMNI PROFILE: NICK—ASSOCIATE TECHNICAL WRITER FOR SUPERCOMPUTER MANUFACTURER

In 2018, Nick graduated from a large public research institution with a BS in Technical Writing and Communication, emphasis in Information Technology, and with minors in Leadership and Digital Media Studies. He particularly enjoyed working with technology and so took computer science and programming courses as part of his emphasis. Reflecting on his decision to take these courses, he suggested that they were really helpful "for laying the framework for understanding how software and other technological components work" and for viewing software from an engineering perspective.

As an undergraduate, Nick completed two internships. His first internship was that of a technical writer for a small electronics and smart technology company. He wrote data pages and quick start guides for various smart home technologies, and he edited manuals for small networking communication devices. His second internship was that of a marketing writer for a non-profit organization. In that role, he helped plan a large event, conducting demographic research, creating marketing materials, and identifying strategies for outreach to people who would attend the event.

Upon graduation, Nick wanted to work in technology, and he landed his current position as an associate technical writer for a supercomputer manufacturer two months after completing his degree. In his position, which he had held for about 18 months at the time of the interview, Nick was the lead writer of a large administrative guide for a software product.

His day-to-day activities included working closely with multiple engineering teams via Scrum meetings (the teams did Agile project management), learning the various software components that he would be documenting in his manuals, testing the procedures that he was working on to ensure they supported user goals, and meeting individually with engineers to plan the best workflow for using different software components. Nick spoke about how each documentation project had a lead writer who created, edited, and published the documentation for the project. They did this using the Oxygen XML Editor and the Astoria Component Content Management system, which supported the DITA content standard. Documentation was published on the company's online portal.

The leadership structure for technical documentation included a technical writing department with two managers, one focused on hardware and the other on software. While the managers oversaw technical documentation projects, individual writers were integrated into engineering teams. This structure, suggested Nick, allowed writers to learn about and keep up with new software developments.

When asked what skills and qualifications are needed to move into more advanced content-focused roles in his organization, Nick talked about project management and leadership skills as well as "having a firm grasp on improving department practices and making sure that our documentation is held to a certain standard." He had ambitions to move into more strategic roles, and he was working to do so by taking on a larger workload and managing more projects on his own. His goal was to get a better feel for what higher-up positions were doing and what it took to succeed in those positions at his company.

So too can entry-level technical communicators expect to work closely with UX teams who collaborate with developers and engineers on product design. Several hiring managers responding to our survey and interviews emphasized the need for UX teams and technical content teams to collaborate, as in-product user assistance is part of the customer journey that connects with other content touchpoints. Entry-level writers often thus help to craft user interface texts such as instructional text, hint text, tooltips, and other embedded user assistance. Because users can now provide immediate feedback on content published on social media, content portals, and community forums, writers also are increasingly expected to review feedback and update content promptly. Structured content enables writers to do this as part of their workflow.

The collaborative environment in which entry-level writers on technical content teams now work also affords them opportunities to develop and showcase new expertise. Advanced-level roles such as content strategists, information architects, data analysts, and structured authoring (including DITA) specialists will continue to emerge as technology and publishing platforms evolve and as the need for approaching content as a discipline evolves. Opportunities for eager, qualified technical writers to move into more advanced roles in content design or new roles in content strategy, content operations, or content engineering will thus only grow. The aging workforce that we discuss in Chapter 4 will also open up opportunities for the eager and qualified.

BOX 3.3 ALUMNI PROFILE: JESSICA—TECHNICAL WRITER FOR FLUID HANDLING EQUIPMENT MANUFACTURER

In 2017, Jessica graduated from a large public research institution with a BS in Technical Writing and Communication, emphasis in Information Technology and Design. She also completed a minor in Interdisciplinary Design. During her senior year, she landed a technical writing internship with a fluid handling equipment manufacturer. Upon graduation, she was hired full time by the company and had worked there for 31 months at the time of the interview. She started as an associate technical writer and, after 22 months, was promoted to an intermediate technical writer.

As a technical writer, her main job was to write and update product manuals for equipment manufactured by her company. She specifically focused on documenting machinery that contractors use at construction sites, for example, to pump concrete or spray foam insulation. Updates were required whenever a new feature was added to a product—she would write a new description and procedure on how to use the feature. Much of this work required her to work closely with SMEs, interviewing them and learning all she could about a product so as to put together an initial manual draft. That draft would then be sent "out with a field test product to an actual customer," and Jessica would then use feedback to finalize the draft. Rather than writers being in one department, they were assigned to different divisions within the company and reported to different engineering managers depending on the division for the type of product, such as products used in medical and food safety environments. Writers were distributed over four locations.

As part of the manual development process, Jessica frequently consulted with her company's legal and safety department, making sure that users were adequately warned about potential hazards and ways that they could avoid them and making sure that her team was using the right trademarks and wording them in a careful, clear way. Jessica also talked about what she referred to as a "side gig" outside her regular job responsibilities. She described helping to develop digital products—or apps—to connect with and control a product remotely as a new task that she was embracing. She was also helping to conduct usability tests for a new distributor portal, a website that her company used to sell products to distributors that sold their products. The site was undergoing a complete redesign and thus frequent end users would need to learn how it worked and how and why it was different from the old site. A course focused on usability testing and human factors that she took as part of her undergraduate major had prepared her well for this kind of work, which she described as "really fulfilling" and fun in that it allowed her to talk with a lot of people.

Jessica talked about a big transition coming to her team. While some writers had been working in DITA for nine years, most were authoring content using "unstructured Adobe FrameMaker." Different authoring approaches had resulted in writers publishing manuals and other deliverables using several different platforms; regardless of platform, the only output enabled by FrameMaker and their DITA-supported systems was that of PDF. This inconsistent and limited approach to technical publications led the company's information architect, a single corporate position in corporate technical publications, to begin conversations with a consultant solutions provider about streamlining authoring approaches and fully adopting DITA in ways that fit the company's publishing needs. Because Jessica had experience with DITA via a writing with technologies course that she had taken for her undergraduate major, she felt comfortable participating in those conversations and helping to plan the transition from a largely unstructured to structured authoring environment. She talked about how the DITA migration needed to be part of a larger "change management" and "project management" plan that was visible across the organization and led by a "driving force champion" who believed in the value of the transition. At the time of her interview, Jessica was not sure when the initiative would launch but was looking forward to taking on a meaningful role in the transition. She spoke, as well, about her hope to move into a more strategic role in the near future. The only strategic role in the current leadership structure was that of the lone information architect, but the transition opened up possibilities for new strategic roles.

Reflecting back on her undergraduate training, Jessica talked about how she wished she had had an opportunity to study project management. A project management skill set would have prepared her well to help lead the transition to DITA—to help build a requirements document, talk with stakeholders, manage change, and advocate for herself, her ideas, and larger changes to the organization. She commented,

> When I look at the DITA migration, how do I talk to someone who is the CEO of the company and convince them that it matters what tool our writers are using? How do I convince management that these are the people we need to get on board to move this forward? That's huge.

For Jessica, project management was a key skill set needed by anyone in a strategic role. It was one that she was working to develop.

Agile Development

The move toward continuous product development and continuous content delivery has impacted technical communicator activities in significant ways. Next to structured authoring, Agile practices were reported in our interview and survey responses as representing the greatest change in the day-to-day work of technical writers, particularly those working in software development.

As we wrote earlier in this chapter, the most common workflow for continuous product development is an Agile model, which is based on four key values articulated in the Manifesto for Agile Software Development (2001):

1. Individuals and interactions over processes and tools
2. Working software over comprehensive documentation
3. Customer collaboration over contract negotiation
4. Responding to change over following a plan.

The Agile model enables software teams to "deliver work in small, but consumable increments, rather than wagering everything on a 'big bang' launch" (simplilearn, 2024). Baehr (2019) offers the following articulation of Agile workflows and their multiple stakeholders:

> In a sense, they are both adaptive and predictive because they begin with a preliminary plan, which adapts as the project encounters change. They focus less on early detailed plans, specifications, or documentation and more on getting started on the product, learning and adapting as the product design develops, and working in teams to check, validate, and test design efficiencies and functions. Equally important, customers or product consumers are invited frequently to interact as team members along with developers, providing input and advice, engaging in testing as the product develops, and suggesting modifications along the way. The production model then develops through feedback and short, iterative (repeated) development cycles—design, develop, test, seek feedback, respond to feedback.
>
> *(p. 7)*

A key change for technical communicators is that they are involved in the process from beginning to end, from gathering requirements to writing user stories and use cases, and they are continually updating and delivering content (primarily structured content, as it affords being modified, delivered, and updated quickly).

Agile models have largely replaced waterfall models of development, which are considered predictive at the cost of being adaptive. As Baehr explains, waterfall models start from a plan and sequentially progress through the five stages of the plan: (1) proposing, (2) analyzing and specifying, (3) implementing, (4) testing and troubleshooting, and (5) producing and maintaining. “Like the assembly line process,” Baehr writes, “waterfall development relies on a strong design that is as viable at the end of production as it was at its beginning”; “If changes occur, workflow (the downward progression of the waterfall) is obstructed” (p. 5). In addition to not being adaptive and responsive to change and customer needs, the waterfall model meant that technical communicators wrote documentation at the end of the development process—and they had to do it fast to meet release deadlines. It also meant that they had to learn how to use the product based on a prototype or specification, and if they made errors in the documentation (often a manual for end users), those errors could not be changed until the next release.

The widespread adoption of Agile practices has meant that entry-level technical writers increasingly plan and manage content development activities in short, quick cycles; work in close collaboration with developers and product owners; and co-author and author content alongside developers and engineers, often using writing tools and docs-as-code writing approaches (which we mentioned in Chapter 2) familiar to developers and engineers. Below, we highlight themes in responses that we collected from hiring managers and industry and academic leaders on how Agile practices are impacting technical communicator activities and roles in tech-focused organizations that have undergone digital content transformations.

- **Plan and manage content development activities in short, quick cycles.** Agile development, with its focus on short, iterative development cycles, has meant that writers need to be involved in product planning and requirements gathering from day one. Entry-level writers work with stakeholders to learn as much as they can about the piece of software under development and start documenting it right away, as it is being developed. Writers need to learn tools and concepts quickly, and they need to be proactive collaborators able to manage their work at a fast pace. Several hiring managers spoke to this change in environment for entry-level writers:

  Program management methodology is now Agile versus waterfall, which gives shorter cycles for content development activities. Entry-level technical writers independently support products and programs versus earlier when they would contribute to programs led by advanced-level writers.

  —*Survey respondent*

In recent times, product development time has reduced drastically. There are no longer any long releases for software products in most companies. Entry-level authors are expected to work with multiple stakeholders within documentation teams as well as with other stakeholders to complete their deliverables on time. This requires them to grasp things at a faster pace and at the same time, collaborate, plan, and manage their work in a fast-paced work environment. In addition, both new and experienced authors are expected to contribute more than just delivering documentation. They will be greatly respected if they can test products to identify defects, suggest improvements to improve quality of products, workflows, and processes.

—*Survey respondent*

Entry-level writers are expected to work as generalists on the team from the start, rather than first working on one type of content before expanding to the next. Moving to Agile software development has also put pressure on faster ramp-up time. There's far less grace period for new writers to get up to speed on tools and concepts. New writers who struggle to engage confidently and proactively with their SMEs (subject matter experts) will struggle to find success in a faster-paced environment.

—*Survey respondent*

BOX 3.4 ALUMNI PROFILE: LANG—STAFF TECHNICAL WRITER FOR CYBERSECURITY SOFTWARE COMPANY

In 2016, Lang graduated from a large public research institution with a BS in Technical and Scientific Communication and a minor in French and Francophone Studies. As required by her major, Lang took several elective engineering and computer science courses as an undergraduate. She also took courses focused on software documentation using the DITA standard, APIs, and writing style. To gain work experience during the summer before her senior year, Lang interned as a technical writer for a government contractor. In that position, she worked with engineers to update existing manuals.

Upon graduation, Lang wanted to work for a technology company in the Bay Area as a technical writer. She applied for several technical writing positions and was soon offered one as a technical writer for a cybersecurity software company that produces software for searching, monitoring, and analyzing machine-generated data via a web-style interface. She worked in this position for 1.5 years before being promoted to a staff technical writer, a position she had held for four months at the time of the interview.

In her current position, Lang works with engineers via Agile development workflows on upcoming releases. She attends daily Sprint meetings where she hears status updates from engineers that help her plan her work. They often pull stories into the Sprint that they then start working on; Lang checks to make sure that she knows what each engineer is working on so that she can act accordingly. In describing her day-to-day activities, Lang commented,

> a lot of it is just monitoring and being aware and tracking engineers' work so that when they merge code or when they open a PR [pull request], I can review the PR...and all the UI text, and I can see how it affects the docs.

If the PR and/or UI text has significant documentation impact, Lang then files a doc ticket for herself and works through updating the documentation for the next upcoming release.

All documentation is published on a homegrown wiki platform that's "basically like a wiki markup environment." The platform allows writers to "jump in anywhere and edit anything at any time." When a software update is released, customers can offer feedback directly via the platform; this allows writers to immediately respond and update the documentation as needed.

In addition to day-to-day, short-term activities, Lang contributes to long-term planning projects. For example, she described how her team's product was merging with another team's product a couple of months out. She was thus working with the documentation writer on the other team, planning the requirements for the new documentation set. Notably, writers at her company work on different product teams and take on the responsibility within their team for writing, managing, and publishing a documentation set. Managers usually oversee up to eight writers. At the time of the interview, her organization employed 40 technical writers who worked for five different managers. These managers reported to two directors.

Lang was promoted from technical writer to staff technical writer after 1.5 years largely because she had proved to her manager that she could successfully handle releases on her own. She had approached her manager asking to take on more responsibility, and her manager agreed that she was ready. As a technical writer, she was "under the wing of someone else," who did "the brunt of the work." As a staff technical writer, she oversaw releases on her own, doing all of the planning and writing.

When asked about the leadership structure of technical writing teams within her organization, Lang described six positions, from entry level to advanced: technical writers, staff technical writers, senior technical writers, principal technical writers, managers, and directors. All writers, regardless of level, report to their manager; the key difference in position level largely comes down to the amount of responsibility and independence. For her to advance to the next level, she would need to be able to independently handle multiple projects with complex features at a time and to be able to create "a whole new documentation set." Moving to this level, she commented, required more experience, more of a system view of products and documentation sets, and a track record of making an impact outside of a direct product area.

- **Work in close collaboration with developers and product owners, most often as a member of an integrated team.** Writers are expected to interact with SMEs such as engineers and developers to help plan projects and develop service and support documentation. The ability to understand product requirements and SME explanations, as well as familiarity with product development processes, is a must. So too is the ability to ask good questions to get the right information and to give and apply feedback. Several hiring managers and leaders spoke about how writers are now embedded in development teams in Agile environments:

  Authors are now more embedded in development teams as part of Agile methodology. Authors have more control over their content, and we have quicker publishing cycles.

  —*Survey respondent*

  I don't know if we talked about Agile at all, but you are now inserted as part of the Scrum (an Agile team collaboration framework) team documenting stuff at each [iteration], ...you don't wait until the end anymore. That is so 1980.

  —*President of a content professionals employment agency*

  So it seems to be that the boundaries are much more fluid than they used to be. We saw far less of that stereotypical, you know, that lone rider in the cubical thing, but we did see that in a couple of places. It was mind-deadening to sit and observe, but there was much more of open environments with a lot of interaction between people. The tech comm folks are key players on projects in the whole Agile process and all of it.

  —*Professor and program head of technical communication*

BOX 3.5 ALUMNI PROFILE: HOLLY—CONTENT DEVELOPER FOR MULTINATIONAL TECHNOLOGY COMPANY

In 2018, Holly graduated with a BA in English from a large public research institution. She majored in Professional and Technical Writing and had minors in Social Change and Leadership. During her senior year, she took courses on user documentation using the DITA standard, technical editing, and web development.

While an undergraduate student, she completed an internship with Recreational Sports on her campus, conducting student research and creating infographics and other documents that communicated to students how they could better themselves through recreational sports. She also completed two marketing internships: one with a winery doing social media strategy and writing articles for their blog and magazine and one with a financial company writing marketing content.

Holly's post-graduation goal was to pursue a career in technical writing. She began applying for jobs the summer before her senior year; she interviewed for a few positions but did not feel they were right for her. In April of her senior year, she applied for a content developer and content designer position at a multinational technology company, well known for its decades of technological innovations; the company provides clients infrastructure, software, and consulting services to support their digital transformations. She was offered the position soon after graduation.

In her position, which she had held for one year at the time of the interview, Holly worked on both internal documentation and external customer-facing documentation for the cloud. She did this in collaboration with support teams and development teams, participating in Scrum meetings and various other daily meetings. Holly described her participation in daily Scrum meetings as "awesome"—at these meetings, writers, designers, and developers would talk about what they did from yesterday to today and what they planned to do in the future. Holly would show everyone the projects that she was working on and explain what she needed to do to make progress.

Holly's team, consisting of five writers, a terminologist, an editor, and a content strategist, was charged with creating and curating documentation for the cloud. They did this using tools that support the creation, management, and delivery of DITA-based, structured content. Holly described the publishing environment as one that gave writers the freedom to do what they needed to do. The terminologist, an editor, and a content strategist had to approve anything that was published, but they trusted writers to figure out what documentation was needed and consult with others on the team and beyond as needed. Upon approval, new or updated content would be published and become accessible to anyone subscribing to the cloud documentation.

When asked what she considered to be more strategic content positions in her organization, Holly mentioned content strategy and terminology. To move into these positions, in her view, writers needed to gain more experience and meet more people in the organization so as to become known and respected for their work. Writers also needed to learn the organization's style and guidelines at a macro level, as those in strategic positions oversaw the organization's documentation as a whole, not "just one specific subsection." Reflecting back on her undergraduate education, Holly felt that she would have been better prepared for her current position and for moving into more strategic positions had she taken more leadership courses and business-oriented courses.

- **Co-author and author alongside developers and engineers, using writing tools and docs-as-code approaches to writing familiar to what developers and engineers use.** In Agile environments, it is common for developers and engineers to contribute to technical content workflows, particularly when documenting APIs. As we have described elsewhere (Evia & Andersen, 2020), "The need to document APIs often resides in a communication structure that has SMEs taking the roles of authors and readers"; SMEs write in accordance with pre-defined templates and standards while writers draw on their extensive knowledge of the product to document code in collaboration with developers (p. 221). An influential factor in SMEs assuming author roles is the need to document products (particularly APIs) for an audience of developers, not just end users. Tom Johnson (2018), an influential blogger and API documentation specialist, has argued that developers and engineers are now required "to play more active, collaborative roles in documentation" because the tools and languages they use "are becoming increasingly specialized and complex" (para. 1).

  SMEs contributing to content workflows as they develop software have also meant a change in writing tools and approaches. Agile practices enable SMEs and writers to treat modular content as computer code, and for developers in particular, writing content as code feels natural and integrates well with development workflows, which can use the same version control system (e.g., GitHub) and repository for both code and content. Known as the docs-as-code movement (where docs means documentation), this approach to content creation, management, and delivery emphasizes the integration of content into the overall development process and often leverages version control systems such as GitHub, collaboration tools, and automated processes commonly used in software development. This approach also requires an authoring format that resembles programming languages, and that format can sometimes be based on XML and DITA but is more frequently represented in a lightweight language such as Markdown reStructuredText or AsciiDoc.

Docs-as-code approaches allow both SME writers and technical writers to seamlessly integrate content (or documentation) into Agile workflows, ensuring that documentation remains an integral and well-maintained part of the software development life cycle. Writers can build, test, and deliver documentation along with the software with both evolving together. Version control systems such as GitHub help writers manage documentation alongside code and ensure that documentation changes are versioned and can be easily rolled back or reviewed. By adopting a docs-as-code approach, SME writers and technical writers can contribute to creating more accurate, up-to-date, and comprehensive documentation, which ultimately benefits both the development team and end users. It streamlines workflows, enhances collaboration, and ensures that documentation is an integral part of the software development life cycle. Hiring managers working in Agile environments emphasized the need for writers to adopt developer processes and tools and to be active participants in the software development life cycle. The following two responses from hiring managers speak to this change for writers:

> There's much more pressure, less time between releases, use of tools that aren't really technical writer tools but are more engineering-focused (so the kinds of tools that engineers think technical writers want). There's also much less focus on quality within these organizations as a whole.
>
> —*Survey respondent*

> Processes and technology changes such as adoption of Agile methodology require writers to adopt those processes and tools. Writers must be able to scope their work using the same tools that development teams use. This requires adoption of epics/stories to scope work, regular sprint planning, retrospectives, and continually attending to the product development cycle. These changes mean that writers must develop their own project management skills and be ready to adopt the workflow of the teams they are a part of.
>
> —*Survey respondent*

Generative AI Impacts on Technical Content Processes

Fears of AI reducing the demand for technical writers or fueling layoffs across the discipline of content have been high since the launch of ChatGPT in November 2022. Technical writing jobs have indeed been impacted by AI, but most of the content leaders we have been following agree that the impact will be less about jobs going away and more about jobs changing to include different foci and roles. In the March 2023 webinar "Is ChatGPT going to take my job?," presented as part of the *Content and Coffee* webinar series, Michael Iantosca, Senior Director of Content Platforms at Avalara, stressed the need to think of large language models as assistive tools (Abel & Bosek, 2023, March 27). He said, "Having assistive technology does not mean that writing jobs are going to go away; it means that writers have more time to focus on higher-level tasks that we haven't had enough resources to do." Automation of repetitive tasks will free up time for writers to focus on writing engaging content, gathering user data to inform journey maps, improving the usability of content deliverables, and other tasks that require human understanding of complex contexts and processes. Those resistant to change and adaptation will be at the greatest risk of losing their jobs.

In a December 2023 episode of *Content + AI*, Lisa Jennings Young, Head of Content Design for Microsoft Teams, shared a similar sentiment (Swanson, 2023a). Using AI to complete repetitive tasks and to help with aspects of content creation frees up time for content practitioners to focus on more impactful work, such as A/B testing, experimentation, more research, journey mapping, and thinking about the content ecosystem. And because AI does not have research insights or an understanding of product strategy, business goals, or user needs, Young believes that generative AI is not going to replace writers. Those who can harness the power of AI, she suggested, "will produce the most outstanding, the most scalable, the most impactful work." Those who are not using AI as an assistive tool are most at risk for being replaced not by AI but by writers and content designers who use AI as an assistive tool.

Tom Johnson and Fabrizio Ferri Benedetti also conversed on this topic in a November 2023 podcast focused on AI and technical communication (Johnson, 2023b, October 20). They agreed that higher order technical writing tasks, particularly around the processes we create to support the creation of deliverables, cannot be automated. The tasks situated in complexity and ambiguity, Johnson argued, such as defining user requirements and figuring out what needs to be written, require understanding and reasoning. Similarly, pondering on generative AI and the future of technical communication, Clara Christopher (2023) emphasized the need for human judgment on both the input and output side of AI and suggested, like Iantosca and Young did, that the biggest impact for technical writing activities and roles will likely be greater focus on higher level tasks and impactful work. Tasks such as streamlining user searches and generating text summaries will be done by AI, but the quality of AI output will always be a "reflection of the team of humans at either end of it"—that is, "the ones who monitor the quality of the input and output."

Given that we are very much in the early adoption stage of generative AI in the discipline of content, we cannot say with any certainty how AI will ultimately impact the activities and roles of technical communicators. But we can highlight what leaders in the content and AI space have so far had to say about how activities and roles are changing now and will likely change in the future as AI is more fully integrated into content workflows. In the following sections, we provide our takeaways from the many webinars, podcasts, whitepapers, blogs, and conference talks that we have watched, listened to, and read on this topic, as well as personal conversations we have had with leaders in the discipline of content. These takeaways center on four key activities that leaders predict will define technical communication work as it relates to AI: (1) creating quality source content, (2) writing in collaboration with AI chatbots, (3) crafting effective prompts, and (4) using AI in all stages of the content life cycle.

- **Creating quality source content.** The old saying "garbage in, garbage out" holds true for generative AI and is a key point being made by all who have contributed to the AI and technical communication conversation thus far: AI systems are only as good as the content that they are trained on or fed—so, if you give them garbage content, you are going to get garbage results. The first point to consider here is the critical role that technical writers play in creating new content for new products—products that have yet to be documented. On a September 2023 episode of *The Content Strategy Experts* podcast, Jack Molisani and Bill Swallow spoke to the absurdity of expecting AI to generate valid technical content (Swallow, 2023, September 11). In defending the need for technical writers, Molisani asserted, "We as a technical profession document things that don't exist yet: that is our reason for existence, and how are you going to get that to an AI?"

  A key activity for technical writers moving forward will be to create the quality source content that LLMs need to be able to generate quality, reliable, and accurate output. Johnson (2023a) has called this emerging activity "writing for AI consumption," similar to what Gallagher (2017) described as writing for algorithmic audiences. The writing principles that will be required to do so, suggested Johnson, include using ample headings and subheadings, semantic tags, code samples, fewer images, chunked information, consistent terms, cross-references, and glossaries. This is because LLMs need quality, searchable source content to be able to answer questions about a company's products. Ellis Pratt, in a November 7, 2023 episode of *The Content Wrangler Webinar Series* focused on generative AI use in TechComm (Abel, 2023a, November 7), cited similar best practices for writing quality source content. At minimum, he explained, these practices will include "using headings and topics, writing clear sentences, using numbered or bulleted lists, avoiding ambiguity, identifying who the content is aimed at, and explaining key terms." These are skills that technical writers bring to the table, and they are skills that will be more, not less, important in an era of AI.

If these practices sound a lot like those required to create structured content, they are, though Johnson and Pratt in this case were describing how writers will need to adapt linear writing processes to improve the ways AI interacts with their content. There's a growing consensus among content leaders, however, that structured content is ideal source content. An AI chatbot connected to a corpus of technical content can better provide user-specific information if that source technical content is written in XML (DITA), JSON, or even consistent Markdown. Pratt (Abel, 2023a, November 7) and others have argued that well-structured, meaningfully tagged content results in better output because semantics tell the system "this thing is a person, this thing is a butterfly, this thing is not a street." On the "AI in Technical Communication and Content Strategy" episode of *Content + AI* with Larry Swanson (2023b), Sarah O'Keefe explained the importance of semantic markup as follows:

> So, if I have a collection of articles and every single one of them has a chunk that's tagged "summary," and then I ask it to give me the most important points from these various articles, I could reasonably expect that the summary would contain those points, and probably I can write a prompt that says, "Look in the summaries," as opposed to processing all of the articles. So, what metadata and semantics...give you the ability to do is essentially provide a roadmap or a set of guardrails or whatever you want to call it to the engines that are processing that content, so that they can more successfully and more accurately reach in and get the information that you want them to get.

Not only is structured content the foundational tool of digital content transformations, but it is also proving to be vital to leveraging the power and potential of AI for meeting both business and customer goals. In fact, in an episode of *Content and Coffee* with Scott Abel on "Real-World Use Cases for Generative AI in TechComm," Casey Jordan, Chief Product Officer and Co-Founder at Heretto, identified three content attributes needed to leverage AI in technical communication (Abel, 2023b, October 4):

- Centralized high-quality content sources
- Modularized content that is well typed (e.g., tasks, concepts, references)—allows AI to traverse and consume the information more effectively
- Enriched content (metadata, semantic content, etc.)—allows the AI to understand what a certain portion of your content is about.

Jordan argued that:

> If you don't have structured content, if your content is not modular, if it is not classified well, if it is not semantic, if there aren't good relationships between your content, [generating quality outputs] gets harder and harder to do really well.

The chatbot will not be able to provide high-quality answers or complete high-quality tasks.

Writing for AI consumption will clearly be a primary activity of technical communicators for some time. Many leaders contributing to content and AI conversations predict that traditional search may soon be replaced with AI chat interfaces that serve as the primary user interface for accessing documentation and relevant content (see, e.g., Abel, 2023b, October 4; Duin & Pedersen, 2023; Johnson, 2023a, April 17; Swanson, 2023b, November 12). The idea is that the chat interface connects with a language model trained on a high-quality corpus of organizational content. When this happens, users can prompt the chatbot with any number of questions or scenarios and receive high-quality output that meets their exact needs for the problem that they are trying to solve.

- **Writing in collaboration with AI chatbots.** How generative AI can be used as an assistant in all stages of a content workflow has perhaps received the most attention in discussions of the impact of AI on technical writing. The future of technical writing no doubt has AI at the center of the writing process, assisting with tasks such as brainstorming, planning, gathering resources, analyzing data and text, researching, summarizing, synthesizing, drafting, revising, editing, and providing feedback. In some cases, technical writers working in organizations that have undergone digital content transformations are using chatbots trained on public data sets, such as ChatGPT or Gemini, to assist in tasks. And with effective prompting, these chatbots can be incredibly productive partners that help writers complete challenging tasks in a fraction of time, such as summarizing a series of reports or analyzing support call transcripts. But for the reasons we discussed earlier, most technical writers can soon expect to be using AI chatbots that have been integrated into their organization's authoring tools that then interact with a large language model (LLM) trained on a specific corpus of content.

Lance Cummings (2023), an associate professor of professional writing who writes about AI and content operations, predicts that writers' approaches to interacting with these AI chatbots will be like a "symbiotic dance"—a back-and-forth exchange that is more like a tango than a transaction. Cummings urges writers to view this AI encounter as a posthumanist moment that asks us to embrace the machine as a collaborator, as an extension of human capabilities. Posthumanism, he suggests, "urges us to re-evaluate the boundaries between human beings and technologies," to see technologies as partners that don't replace human creativity but rather augment it. In a posthumanist view, "writing becomes a networked, interactive endeavor" where "human and machine input feed into each other." This interactive dance is how content leaders imagine writers collaborating with AI in all the work they do. As Urbina described in his conversation with Swanson (2023c, November 8) on the *Content + AI* podcast, "you can iteratively go back and forth, draft some content, you edit it, you improve it, send it back, ask for improvements." Through feeding chatbots well-defined inputs and expected outputs, writers will engage in collaborative exchanges that result in high-quality products in far less time than would be the case without the chatbot writing assistant.

Many companies have embraced AI as an augmentation tool in the writing process. Microsoft, for example, explicitly encourages its writers to use generative AI for technical content published on Microsoft Learn. Technical writers use Writer, a full-stack, integrated platform for enterprise generative AI, to augment their content-generation process and then review and revise AI-generated content. Articles must pass a validation test, a tool that checks, among other things, for "formatting errors, and to make sure the terms and language are appropriate and inclusive." Once approved, articles are clearly marked as containing AI-generated content. Disclosure of use of AI is paramount in

Microsoft's policy around using generative AI for technical content, as articulated on its Microsoft Learn page, "Our principles for using AI-generated content on Microsoft Learn" (Microsoft, 2024). Guardrails in place ensure that published content is accurate and audience aware and transparent about the extent to which AI augmented the generation process.

Microsoft's approach to using AI as an augmentation tool speaks to the importance of having humans in the loop (a key point Tham, Howard, and Verhulsdonck make in their discussion of AI and content automation, 2024). The more AI is integrated into content workflows as a collaborative partner, the more writers will need to assume the role of critical editors at each stage of the process. Writers will be responsible for reviewing AI output for accuracy, completeness, consistency, coherence, brand compliance, and overall effective communication of the message.

We are aware that this space is advancing exponentially and that by the time this book is published many of these examples will be obsolete. However, the AI chatbot Etto, released in late 2023 by the content operations platform Heretto, is an early powerful example of how we can expect to see AI integrated into structured authoring tools and workflows in the near future. Heretto was an early adopter of this technology as it intersects with technical communication and the discipline of content. Knowing that many content contributors using the system were not experts in creating structured content, Heretto (2023) wanted to make the ability to create structured content as easy as possible to help contributors create "exceptional content experiences" (p. 2). Content contributors engage with Etto, a DITA "expert," as they author structured content; Etto answers questions, offers best practices, creates and revises content, and self-identifies when it produces subpar responses. Example DITA-compliant tasks that Etto can do include creating new content, such as inserting a short description in a concept topic or converting an unstructured procedure into a task topic; improving and updating content, such as updating a section title or steps in a task based on text from a support ticket; and providing guidance, such as instructing a writer in how to fix a validation error or how to semantically enrich a topic.

AI chatbots like Etto hold the promise of democratizing structured authoring, making it easier for all content stakeholders, including SMEs and product managers, to create structured content. It is not yet clear how chatbots trained as experts in DITA, JSON, Markdown, and other authoring languages will impact technical communicator roles, but we can safely assume that they will not replace writers or the need for writers to understand key concepts of and approaches to structured authoring.

- **Crafting effective prompts.** To generate quality output, AI chatbots first must have access to quality source content, the creation of which will be a primary activity for technical communicators in the future. AI chatbots, as of mid-2024, must be given clear instructions in the form of a prompt, or the initial input provided to a language model to generate a response. A prompt, as defined by Ellis Pratt (Abel, 2023a, November 7), is written in natural language and "helps the model understand the user's intent and provide a meaningful response based on the information it has been trained on."

  Many prominent voices in AI conversations (e.g., Fabrizio Ferri Benedetti, Tom Johnson, Anthony Medeiros, Noz Urbina) predict that some technical communicators will move into prompt engineering roles, where their job will be to "engineer" the task that they or other stakeholders want the AI chatbot to do. Urbina, in his conversation with Swanson (2023c), equated the prompt engineering role to that of programmers. Whereas writers used to rely on those who could code to explain to the computer what they wanted, now writers can write the code in the form of prompting because writers understand what they want. Urbina sees great potential for this new role for those who embrace it:

  > AI enables us all to be programmers without being coders. Those who can best understand the problem, its context, its constraints, the intended results, and are able to express that logically and clearly will win the future.

  Technical writers have long been trained to think and program in this way. They are trained in rhetorical analysis, clear writing strategies, audience awareness, and task-based writing. Writers who become skilled in adapting this training for engineering prompts will not only contribute value to their teams but open doors for advancing into more strategic roles that support AI integrations in content operations.

  Crafting effective prompts means knowing how to structure prompts. Cummings has spoken and written extensively on the topic of structuring AI prompts (see, e.g., Abel, 2023c, 2023d and Cumming's blog iSophist). He argues that generative AI requires content professionals to be explicit about how they frame their content, about how they structure prompts and define rules for chatbots to follow. Use of generative AI in organizational contexts also requires content professionals to think about AI operations and the need for prompt consistency and prompt management. He thus advocates for applying structured writing principles to prompt creation, as generative AI chatbots recognize structure in inputs and prompts and produce the best results when they understand what specifically they are being asked to do. Cummings (Abel, 2023c) described the need for teams to adopt a shared structured approach to prompting AI as follows:

> Structured prompts are crucial for teams and organizations seeking consistency, accuracy, and efficiency in their AI operations. If everyone is prompting AI in different ways, you're going to get all sorts of different content. Not consistent content.

The idea here is for teams to share a prompt repository of sorts, with prompts being chunked into component parts based on their purpose and function, in much the same way DITA classifies topics into concepts, tasks, references, and so on. These component parts are then saved as individual files and can be called up by prompters to populate a prompt.

In 2022 and 2023, conversations among leaders in the content and AI space focused on strategies for engineering prompts to leverage the power of AI. Example strategies, among others, included creating "prompt blocks" or "predefined sets of instructions given to the AI to guide its responses" (Cummings, in Abel, 2023d, September 19) and using elements, such as [Role], [Context], and [Request], that serve different functions and communicate specific instructions to the AI chatbot (Kaufman, in Abel, 2023e, Aug. 8; Medeiros, in Abel, 2023f, August 15). Webinars such as "Structuring AI Prompts in Technical Communication" (Abel, 2023d, September 19) and "Unlocking the Power of AI in Content Creation" (Abel, 2023e, August 8) gained a lot of traction among technical writers looking to develop a prompt engineering skill set. At that time, effective prompt engineering was also promoted as a way to minimize bias and hallucinations in the output of AI chatbots trained on public data sets (Urbina, in Swanson, 2023c, November 8).

At the same time, influential authors in the AI sphere are calling for users to interact with large language models as if they were collaborators and not software products. GenAI models are programmed to perform like people, and their exponential development is advancing rapidly in this direction and away from the need for specific prompt languages. For example, Ethan Mollick (2024), an academic who writes about innovation and has become a prominent voice in conversations about AI, calls prompt engineering a likely useful skill in the near term but argues that it won't be necessary as language models evolve.

Within the next year, as this book will go through the publishing process, we expect to see a plethora of resources on engineering prompts for generative AI. All technical writers will be expected to converse with generative AI through carefully crafted inputs, and many technical writers will move into specialized prompt engineer roles, focusing on taxonomy development, metadata strategies, prompt management, and more.

- **Using AI in all stages of the content life cycle.** If they are not already doing so, technical writers will soon be using AI in all stages of the content life cycle. As of late 2023, leaders were just beginning to learn how to leverage AI for research, customer journey mapping, data extraction, structuring authoring, editing, content delivery, and more. At that time, technical communicators' use of and thinking about AI had been limited, a point Urbina emphasized in his conversation with Larry Swanson (2023c):

  > There's all these different touch points in an omnichannel experience. Where could AI help each of those as opposed to just always focusing on that, okay, people are going to go into a chatbot and then they'll get an answer. That's a drastically over-limited way of thinking about the benefits of this technology.

  Table 3.2 presents a brief overview of other emerging use cases that content leaders predict will be part of the content professional's workflow in the near future. These use cases have been discussed in podcasts and webinars featuring Noz Urbina (Swanson, 2023c, November 8), Ellis Pratt (Abel, 2023a, November 7), Jack Molisani (Swallow, 2023, September 11), and Casey Jordan (Abel, 2023b, October 4).

While yet to come to fruition, chat-based user interfaces are largely predicted to change how technical content is delivered, replacing online help and documentation portals (see, e.g., Abel, 2023b, October 4; Duin & Pedersen, 2023; Johnson, 2023a, April 17; Swanson, 2023b). O'Keefe, in her conversation with Swanson (2023b) on an episode of the *AI + Content* podcast, explained the chat-based user interface as an information access point this way:

> So, downstream the end consumer instead of resorting to search is resorting to, I'm going to use this chat interface because this feels like a more conversational way of handling it. And ideally, in a scenario like that, we really want them to rely not on the internet, the large learning model, the enormous amount of content that's sitting inside that chat interface, but rather on the vetted, approved, and probably restricted

TABLE 3.2 Overview of Emerging Use Cases for AI Use in the Content Life Cycle

Sample Use Cases	
Transcribing and synthesizing research	AI can transcribe recorded interviews, support calls, sales calls, and chatbot logs and synthesize, structure, and analyze them. AI can also do analysis of data sets, summaries of research reports and articles, and many other tedious research tasks.
Creating personas and journey maps	AI can be trained to be a particular persona and then respond as that persona at each stage of their customer journey, identifying problems encountered, tasks they are trying to complete, questions they have on their mind at a particular moment, and more. AI can fill in a journey map as it is guided through each stage of the customer journey.
Extracting data from files	Writers can use AI to extract information, such as specifications, from documents created by SMEs, business analysts, and others who have information that writers need. Writers can then ask AI to format the information for particular uses, such as a DITA table in a product description.
Converting transcripts into instructions	Writers can use AI to transcribe product interviews with SMEs and then ask AI to convert the transcript into structured topics. The AI will need extensive training to do this, and the output will need cleaning up, but the efficiency gains will be a major motivation for using AI to assist translating complex information into usable and useful content.
Personalizing user assistance	Writers can create structured content wrapped in semantic tags that an AI bot can then interact with in response to a user request. A backend system gets information about the user, and then via a simple user interface the user enters information via drop downs that provide the context the AI bot needs. The user then asks the AI bot a question and the bot returns a personalized answer.
Automating tasks	AI can automate a plethora of tasks, from completing repetitive, routine tasks such as extracting data from files and populating tags to reading hundreds of topics in a CMS and identifying which ones are similar and thus prospects for combining and reusing.
Explaining concepts, processes, code, and more	Writers can use AI for general assistance, such as explaining a technical concept, product feature, or line of code. AI can also be a productive brainstorming partner, helping writers plan projects and approaches to documenting products.
Assisting in creating structured content	AI bots trained on DITA, Markdown, JSON, and other languages used for authoring structured content will soon be integrated into all authoring tools. Some of this is already available in tools like GitHub Copilot. AI will thus help writers convert unstructured content into structured content, validate modular topics, update content elements such as short descriptions, and much more.

> content for a particular topic. In other words, if I make a particular set of products, I want you to rely on that database not on the generalized generic big AI stuff. I want you to process that smaller, more targeted set of information to look for answers.

Johnson (2023a, April 17) speculated that AI chat interfaces will become the primary user interface to read documentation. Quality source content will be imperative, securing the need for content professionals skilled in writing structured content and engineering context-aware, structured prompts. Documentation, suggested Johnson, will become "the source that powers the AI chat interface" and thus "must be written for machine consumption." What will matter is not "so much the UX of your doc site, but the UX of the AI chat based on the information it consumed from your documentation." Because quality source content is required for training the language model, the need for documentation will only grow. Johnson further suggested that the

> more documentation a company can produce, the better AI will be. And the better the AI, the more likely users will use your product and succeed with it. This only suggests that documentation will play a greater role in the future.

AI chat interfaces have the potential to drive higher customer engagement and produce better analytics on customer behavior. Better analytics would allow content professionals to better understand what customers are struggling to find, what product features are important to customers, how challenging content is to understand, and more. This understanding then would enable content professionals to improve and better customize source content accessed via AI interfaces at different stages in a customer journey. If indeed AI chat interfaces become the primary user interface to read documentation, technical communicators, suggested Duin and Pedersen (2023), will see their roles change from "solving problems for users to assisting people in solving problems for themselves" (p. 116). They will do this through creating content for AI consumption, with the AI interface serving as the user's self-help partner and learning from the user as they engage with the interface.

While we don't yet know what the future of AI in technical communication will look like, we do know that AI and content professionals will partner in all aspects of the discipline of content. Students preparing for careers in the intersection of technical communication and the discipline of content will thus need both critical and functional AI literacy.

Looking Ahead

The environments in which technical communicators work have changed considerably in the past ten years. The move toward Agile coupled with reduced product development time, the adoption of more complex systems and tools to support digital content transformations, and the embrace of structured content as the foundational tool of digital transformations—all of these changes have significantly impacted the roles and activities of technical communicators in both entry-level and advanced roles. To thrive in organizations that have adopted workflows and technologies that treat content as computable data, technical communicators must be flexible and adaptable and be always ready for change and for learning new skills, tools, and approaches to creating, managing, and delivering content. They must be systems thinkers eager to collaboratively problem solve and contribute solutions.

Rhetorical agency has moved from the model of a lone writer to the interdisciplinary team charged with designing and overseeing content ecosystems. This change requires technical communicators to work and think at a system level that engages stakeholders from across an organization. And it requires them to apply technical expertise to both understand complex products and systems and contribute meaningfully to content ecosystems. Entry-level technical communicators in content-focused roles who aim to advocate for users and move into more strategic roles in the areas of content strategy, content engineering, and content operations must demonstrate a systems approach to thinking about/creating content. They need to be able to articulate where they fit in the system and how the content they create fits into the system. Strategic roles in all areas of the discipline of content tend to require substantial technical expertise; in-depth understanding of business goals, processes, and content needs; in-depth understanding of the product life cycle; and advanced audience analysis skills (e.g., the ability to collect data on and analyze different markets, demographics, and experiences). While it is not realistic for recent graduates to be hired into these more advanced roles, students who are trained to think business-wide and cross-team and in terms of systems and abstractions, structure, and semantics will be best equipped for advancement.

Given the changes in the roles and activities discussed in this chapter, what skills and qualities do hiring managers and key voices in the content industry say are most needed for entry-level content-focused positions? What challenges do hiring managers face when hiring for particular roles and skills sets? What strengths and deficiencies have hiring managers observed new college graduates bringing to their organizations? In Chapter 4, we turn to answering these questions.

Notes

1 In our research protocol, we promised to keep interviewees anonymous in reports. Therefore, we do not disclose their names in this book.
2 https://github.com/VT-Evia/content-discipline.

References

Abel, S. (2018a, December 13). *The state of technical communication: 2019* [Webinar]. The Content Wrangler. Retrieved from https://www.brighttalk.com/webcast/9273/338293.

Abel, S. (2018b). Survey reveals top tools, trends, and technologies in use in technical communication teams. *Intercom*, 33–38.

Abel, S. (Host). (2023a, November 7). Using ChatGPT and gen AI in TechComm: Elias Pratt discusses research survey results, insights, and practical applications [Webinar]. In *The Content Wrangler Webinar Series*. Heretto and The Content Wrangler. Retrieved from https://www.brighttalk.com/webcast/9273/597514.

Abel, S. (Host). (2023b, October 4). Real-world use cases for generative AI in TechComm, with Casey Jordan [Webinar]. In *Coffee and Content*. The Content Wrangler and Heretto. Retrieved from https://www.brighttalk.com/webcast/9273/593704.

Abel, S. (Host). (2023c, August 17). The impact of generative AI on writers and writing: Lance Cummings on how AI is changing the way writers write and the way teachers teach writing [Webinar]. In *Coffee and Content*. The Content Wrangler and Heretto. Retrieved from https://www.brighttalk.com/webcast/9273/588357.

Abel, S. (Host). (2023d, September 19). Structuring AI prompts for technical communication: Lance Cummings on writing clear instructions designed to give you better, more specific output. In *The Content Wrangler Webinar Series*. Heretto and The Content Wrangler. Retrieved from https://www.brighttalk.com/webcast/9273/557984.

Abel, S., & Bosek, P. (Hosts). (2023, March 27). Is ChatGPT going to take my job? Michael Iantosca on practical uses of language models and generative AI in technical documentation [Webinar]. In *Coffee and Content*. The Content Wrangler and Heretto. Retrieved from https://www.brighttalk.com/webcast/9273/577257.

Adobe. (2018). Adobe tech comm survey 2017–2018 findings. Retrieved June 19, 2024, from https://www.adobe.com/products/technicalcommunicationsuite/whitepaper-technical-communication-survey-findings-2018.html.

Adobe. (2022). Adobe technical communication industry survey 2022: Current and future state--key findings. Retrieved from https://www.adobe.com/products/technicalcommunicationsuite/whitepaper-technical-communication-survey-findings-2022.html.

Andersen, R. (2014). Rhetorical work in the age of content management: Implications for the field of technical communication. *Journal of Business and Technical Communication*, *28*(2), 115–157.

Baehr, C. (2019). *The agile communicator: Principles and practices in technical communication* (3rd ed.). Dubuque, IA: Kendall Hunt Publishing.

Bellamy, L., Carey, M., & Schlotfeldt, J. (2011). *DITA best practices: A roadmap for writing, editing, and architecting in DITA*. Upper Saddle River, NJ: IBM Press.

Carliner, S., & Chen, Y. (2018). What technical communicators do. *Intercom*, 13–16.

Carroll, J. M. (1990). *The Nurnberg funnel: Designing minimalist instruction for practical computer skill*. Cambridge: MIT Press.

Center for Information-Development Management (CIDM), the Data Conversion Laboratory (DCL), and the RWS Group (2022, August 25). *Digital transformation survey 2022: Results, analysis, and projections* [Webinar]. Retrieved June 19, 2024, from https://www.infomanagementcenter.com/product/digital-transformation-survey-2022/.

Cleary, Y. (2021). *The profession and practice of technical communication*. New York: Routledge.

Christopher, C. (2023, July 18). Generative AI and the future of technical writing [Blog post]. Security Bloggers Network. Retrieved from https://securityboulevard.com/2023/07/generative-ai-and-the-future-of-technical-writing/.

Cummings, L. (2023). AI and the future of writing – A recap [Blog post]. Retrieved from https://www.isophist.com/p/ai-and-the-future-of-writing-a-recap.

Darwin Information Typing Architecture (DITA) Version 1.3 Part 3: All-Inclusive Edition (2018, June 19). Retrieved from https://docs.oasis-open.org/dita/dita/v1.3/dita-v1.3-part3-all-inclusive.html.

Day, D., Priestley, M., & Schell, D. (2001, March 1). Introduction to the Darwin Information Typing Architecture. Retrieved from https://people.cs.vt.edu/~kafura/CS6604/Papers/Darwin-Information-Typing-Architecture.pdf.

DITAWriter. (2016, July 15). Don Day and Michael Priestly [sic] on the beginnings of DITA: Part 2. Retrieved June 19, 2024, from https://www.ditawriter.com/don-day-and-michael-priestly-on-the-beginnings-of-dita-part-2/.

Duin, A.H., & Pedersen, I. (2023). Augmentation Technologies and Artificial Intelligence in Technical Communication: Designing Ethical Futures (1st ed.). New York: Routledge.

Evia, C. (2018). *Creating intelligent content with Lightweight DITA*. New York: Routledge.

Evia, C., & Andersen, R. (2020). Afterword: Beyond management: Understanding the many forces that shape content today. In T. Bridgeford (Ed.), *Teaching content management in technical and professional communication* (pp. 213–231). New York: Routledge.

Gallagher, J. R. (2017). Writing for algorithmic audiences. *Computers and Composition*, *45*, 25–35.

Hackos, J. (2011). Introduction to DITA: A user guide to the Darwin Information Typing Architecture including DITA 1.2 (2nd ed.). Denver, CO: Comtech Services, Inc.

Heretto. (2023). *Structured content unleashed: Heretto's generative AI breakthrough for technical documentation* [White paper]. Retrieved from https://www.heretto.com/white-papers/structured-content-unleashed.

Iantosca, M. (2025, March 3). DITA—The industry that almost wasn't. Retrieved June 23, 2025 from https://thinkingdocumentation.com/blog/f/dita-%E2%80%94-the-industry-that-almost-wasn%E2%80%99t

Johnson, T. (2018, October 2). Tech comm trends: Providing value as a generalist in a sea of specialists (Part I). *I'd Rather be Writing*. Retrieved from https://idratherbewriting.com/2018/10/02/providing-value-as-generalists-in-specialist-contexts-part-1/.

Johnson, T. (2023a, April 17). AI chat interfaces could become the primary user interface to read documentation [Blog post]. I'd Rather Be Writing. Retrieved from https://idratherbewriting.com/blog/ai-chat-interfaces-are-the-new-user-interface-for-docs.

Johnson, T. (Host). (2023b, October 20). Chatting about AI trends and tech comm with Fabrizio Ferri Benedetti (No. 298) [Video podcast episode]. In *I'd Rather Be Writing*. I'd Rather Be Writing. Retrieved from https://www.youtube.com/watch?v=YyAtNrJQu-0&t=2s.

Kimball, M. (2016). The golden age of technical communication. *Journal of Technical Writing and Communication*, *47*(3), 330–358.

Kimber, E. (2012). *DITA for practitioners, volume 1: Architecture and technology*. Laguna Hills, CA: XML Press.

Lanier, C. (2018). Toward understanding important workplace issues for technical communicators. *Technical Communication*, *65*(1), 66–84.

Manifesto for Agile software development (2001). Retrieved from https://agilemanifesto.org/.

Microsoft. (2024). Our principles for using AI-generated content on Microsoft Learn. Retrieved from https://learn.microsoft.com/en-us/principles-for-ai-generated-content.

Mollick, E. (2024). *Co-intelligence: Living and working with AI*. New York: Portfolio.

OASIS Darwin Information Typing Architecture (DITA) TC. (2018). Retrieved from https://www.oasis-open.org/committees/tc_home.php?wg_abbrev=dita.

O'Keefe, S., & Pringle, A. (2017, April 12). *Structured authoring and XML*. [White paper]. Research Triangle Park, NC: Scriptorium Publishing Services. Retrieved May 17, 2019, from www.scriptorium.com/structure.pdf.

Simplilearn. (2024, May 21). What is Agile? Understanding Agile methodology and principles in software development. Retrieved from https://www.simplilearn.com/tutorials/agile-scrum-tutorial/what-is-agile.

Swallow, B. (Host). (2023, September 11). Adapt to evolving content careers with guest Jack Molisani (No. 151) [Audio podcast episode]. In *The Content Strategy Experts*. Scriptorium. Retrieved from https://www.scriptorium.com/2023/09/adapt-to-evolving-content-careers-with-guest-jack-molisani/.

Swanson, L. (Host). (2023a, December 7). Lisa Jennings Young: Pioneering AI in content design operations – episode 8 (No. 8) [Audio podcast episode]. In *Content + AI*. Larry Swanson and Elless Media. Retrieved from https://contentandai.com/lisa-jennings-young/.

Swanson, L. (Host). (2023b, November 12). Sarah O'Keefe: AI in technical communication and content strategy – episode 4 (No. 4) [Audio podcast episode]. In *Content + AI*. Larry Swanson and Elless Media. Retrieved from https://contentandai.com/sarah-okeefe/.

Swanson, L. (Host). (2023c, November 8). Noz Urbina: The RAUS Method for accelerating content projects with AI – episode 3 (No. 3) [Audio podcast episode]. In *Content + AI*. Larry Swanson and Elless Media. Retrieved from https://contentandai.com/noz-urbina/.

Swarts, J. (2020). Writing about structure in DITA. In T. Bridgeford (Ed.), *Teaching content management in technical and professional communication* (pp. 155–173). New York: Routledge.

Tham, J., Howard, T., & Verhulsdonck, G. (2024). *UX writing: Designing user-centered content*. New York: Routledge.

4

HIRING NEEDS FOR ENTRY-LEVEL CONTENT-FOCUSED JOBS

Conversations on an Aging Workforce and the Need for a New Generation of Technical Content Professionals

The discipline of content for some time now has been abuzz with questions about who is going to take over digital transformation initiatives and more complex roles when the current generation of leaders and innovators retires. The narrative among many experts in this intersection of technical communication and the discipline of content is that a new generation of technical content professionals is needed, but qualified and interested recent college graduates are hard to find. One prominent technical content solutions consultant who regularly presents at conferences such as LavaCon and the (now former) Society for Technical Communication Summit told us that he has noticed in attending those conferences that those regarded as leaders—people defining best practices and planning for the future of content—are all over 45 years old. Others have acknowledged the aging workforce as well, hinting at a potential hiring crisis given the impending wave of retirees. In an interview, a senior documentation manager who had worked in the industry for 30+ years shared the following:

> So this is a huge bulge in the snake, which is all of the 50, 60, 70-year-old tech writers who, at some point, are going to retire. They're retiring now, right, left, and center. We do not know how to replace them.

Results of the 2018 census on the state of the technical communication industry support these observations. Saul Carliner and Yuan Chen (2018a)

DOI: 10.4324/9781003451730-7

found that 48% of census participants were 50 or older, with the largest age group being 56–60. The authors worried about the impact of this finding on the health of the profession:

> On one hand, this aligns with current workforce projections, which suggest that the largest growing segments of the workforce are 50 or older. On the other hand, the low percentage of workers under the age of 35 raises concerns about the long-term status of the profession.
>
> *(p. 9)*

The aging workforce of technical content professionals has prompted many in the industry to strategize on how to attract and prepare a new generation of professionals for the content roles of today and tomorrow. This was a central topic in a January 2022 episode of the podcast *Coffee & Content* (Abel & Bosek, 2022) and in an October 2022 episode of the podcast *Content Components* (Bosek, 2022). We were invited to speak about what we had learned from our research, and particularly from hiring managers, about hiring needs for entry-level content-focused positions and challenges around finding qualified candidates. Several questions from the moderators aimed to tease out reasons why college students do not seem interested in pursuing content-focused careers, an assumption stemming from the relatively few applications that hiring managers were seeing from new college graduates for entry-level "technical writer" or "user assistance developer" positions.

In those podcast interviews, we discussed what we see as a two-pronged problem: (1) college students are not broadly aware that there is a discipline focused on the intersection of technical communication and content, and (2) most college students are not being prepared to work in that disciplinary fusion.

- Most high schoolers have never heard of the field of technical communication, let alone a thing like technical content that sounds like YouTube entertaining videos but it is more about products that enable users to perform specific tasks. When high schoolers begin their undergraduate journey, they are required to declare a major, before they know anything about technical communication (with some exceptions, of course). Those who want to study texts, writing, and/or communication, then, often declare majors that they recognize, such as English or Communication. Of the 2,800+ four-year colleges in the United States, fewer than 10% offer full-fledged or emphasis degrees in technical and professional communication (TPC) (or some variation of the name), and only a handful of these offer curricula focused on the content-focused sector of the field. As a result, even if a college student learns about the field of technical communication at some point in their undergraduate years, they may not have opportunities to study technical communication through their college coursework. We include ourselves as part of the problem and find it humbling to acknowledge that neither industry professionals nor academics teaching in technical communication undergraduate programs have done a good job getting the word out to high schoolers and college students that there is a field called technical communication and that there are robust career advancement opportunities for those who are prepared for entry-level positions as technical content professionals.

- Preparation also factors into perceptions among our industry peers that college students seem uninterested in pursuing technical content-focused careers. With such diversity in technical communication curricula, students potentially interested in these careers do not necessarily get training in the disciplinary areas that connect content to technical communication, including the area of content design, which we see as encompassing technical writing roles and activities. They thus may not be aware of the different career paths available and may not be qualified for the positions that they could see advertised. The result is what our industry peers identify as a dearth in applications for content-focused positions from college graduates with degrees in technical communication and related fields. Several of the alumni whom we interviewed felt unqualified for entry-level, content-focused technical writing positions and therefore did not apply (we discuss the job search experience of eight alumni of technical communication undergraduate programs later in this chapter).

Another piece to the puzzle is the changing nature of the role of the technical content professional. In the *Coffee & Content* episode, moderator Patrick Bosek pondered whether the hybrid nature of the evolving role presented a challenge for college graduates wanting to pursue writing for technology as a career. He commented,

> You have to have some of the writing, but you increasingly have to have some of the actual technical and developer skills. In the past, technical writers...were largely people who worked in publishing and who interfaced directly with engineers. But now there is this whole thing called the content supply chain.

Bosek then posed the question, "Has the industry been changing so quickly that the education side of things has not had a chance to catch up?" In the

Content Components episode that aired nine months later, Bosek suggested that the wide range of approaches in industry to building content deliverables has made it difficult for anyone working in the profession to identify a unifying theme. Agreeing with Bosek, Carlos pointed out that when everyone was producing computer manuals, both industry professionals and academic programs knew what the common deliverable was across the industry. Now, neither industry nor academic programs have agreement on what contemporary transferable skills should be.

The 2022 Center for Information-Development Management (CIDM) Best Practices Conference also dedicated a half-day session to the topic of the aging workforce of technical content professionals and the need to increase engagement in the content-focused sector of the field. Given the focus of our research, we were invited to facilitate the session "Building a Community Table" and speak to the questions, "How do we attract qualified and enthusiastic people to our industry table?" and "How do we remove barriers to entry for those who are interested?" Session participants, all of whom were in management positions, expressed concerns about the number of people planning to retire from their organization in the next five to ten years. They knew they needed a plan for hiring long term, for hiring now for needs in five to seven years. However, given the complex and fast-paced systems within which most participants were working, they also knew that less experienced hires would need ample time to learn on the job and develop the requisite skills and experience for taking over particular roles. Most of their organizations, though, did not have the resources necessary to support training at this level.

Some participants in that session had hired graduates just out of college with excellent results. Others, however, had struggled to find the right skills and experience and questioned whether college graduates were even interested in content-focused careers. There was a clear consensus that the industry needed to find ways to better promote the industry side of technical content and prepare college students and recent graduates for different career paths in this profession.

Clearly, both the industry and education components of the discipline of technical content have much work to do to align hiring practices and curricular pathways so that students can gain the competencies and experience now required for entry-level content-focused positions.

The Rise of Degree-Granting Technical and Professional Communication Programs

At the same time that hiring managers are struggling to find qualified recent college graduates for entry-level content-focused positions, the number of students graduating with undergraduate degrees in TPC is on the rise. Between 2005 and 2012, the number of undergraduate degree programs offering a full-fledged major (n = 65) or concentration (n = 120) in TPC increased by 131% (see Melonçon & Henschell, 2013). In her more recent research on degree programs, conducted as part of her TechComm Programmatic Central Project, Lisa Melonçon (2021) identified 80 programs that offer full-fledged majors (as of 2021) and 225 programs that offer a concentration (as of 2018), representing yet another significant increase. Included in this count are majors and concentrations with variant names, from TPC (most common) to technical and scientific communication to professional writing and rhetoric. These numbers did not include programs specializing in journalism, creative writing, literature, or language arts.

We see several factors accounting for why hiring managers may be struggling to fill entry-level positions despite the growing number of graduates with undergraduate degrees in the field.

1 **Leaders and managers working in the disciplinary area of technical content are often unaware that full-fledged undergraduate degree programs in technical communication (and related variants) exist in the United States.** As a result of this disconnect, many technical communicators in industry do not actively recruit from academic programs or have a good sense of what skills and competencies candidates with degrees in the field can be expected to bring to content teams. Given that few leaders and managers themselves have degrees in the field, undergraduate or graduate, it is not surprising that they may know relatively little about the landscape of academic education in technical communication. Of the 676 technical communicators who responded to the 2018 census on the state of the technical communication industry, only 32% had completed degrees in the field, most coming to the profession after having pursued another career (Carliner & Chen, 2018a, p. 9). Rebekka and JoAnn Hackos (Andersen & Hackos, 2021), in their survey of 187 technical communicators, found the number to be even lower, with only 6% of respondents (n = 9) who had completed a bachelor's degree (n=147) indicating that they had completed their degree in technical communication (notably, 14% indicated that they had completed a master's degree in the field). Both surveys found that the most common area of undergraduate study among respondents was some form of communication and writing, in English, journalism, creative writing, or language arts.
2 **The diversity and expansiveness of work and research in the field are reflected in its undergraduate programs.** Technical communicators work in an array of industries, from information technology services and solutions to manufacturing to government and military to finance and

banking (Carliner & Chen, 2018a). College graduates can thus pursue an array of career opportunities spanning all facets of the field. Opportunities, among many others, include career paths as writers in government agencies, non-profit organizations, or organizations that have adopted digital content transformations, as well as paths as instructional designers, UX researchers, social media strategists, technology trainers, and more.

Most websites of academic programs in technical communication and related fields highlight these varied career opportunities and the many specializations that students might pursue, from web theory and design to scientific and medical communication to technical editing. Boise State University, which offers a major in Writing, Rhetoric, and Technical Communication, for example, allows students to build their own degree based on their career goals. Likewise, James Madison University, which offers both a BA and BS major in Writing, Rhetoric, and Technical Communication, allows students to design their own career track or to pursue a specialized track in areas such as professional writing and publishing, inclusive design, and writing for science and health.

To help students focus on a particular career specialization, many programs offer dedicated curricular pathways (i.e., a sequence of courses that students can take). Across programs, these curricular pathways may share similar kinds of courses and learning outcomes (e.g., scientific communication, UX, editing, and publishing); however, the field as a whole does not standardize curriculum. What this means is that a technical writing concentration in one university may look very different from the same concentration in another, and thus students graduate from those programs with different sets of competencies, skills, and experiences. As of the writing of this book, only a handful of undergraduate programs offer a technical communication/writing concentration grounded in approaches to technical content.

The diversity of programs and the concentrations that they offer, including the variety in curricular approaches to preparing students for particular career specializations, has afforded graduates the opportunity to pursue a range of careers. But it also has meant that most graduates come from programs that do not offer curricular pathways for technical content-focused positions. We suspect that graduates who specialized in technical writing, in particular, but who did not complete a curricular pathway grounded in aspects of the discipline of technical content, may see job ads for entry-level content-focused positions and feel discouraged to apply. Or, they may be more interested in applying for traditional technical writing positions in government or other sectors of industry and services.

- **The diverse foci and curricular structure of educational programs reflect faculty strengths and the local needs and constraints of institutions.** Faculty who teach in technical communication programs come from a wide range of backgrounds, many having completed doctoral degrees in technical communication or in rhetoric and composition but just as many having completed graduate degrees in other fields, such as literature, linguistics, or information science. Some have extensive workplace experience and some have none, and some teach part time while working full time in industry (a handful of these are content professionals in leadership roles). Technical communication curricula across US institutions reflect the diverse backgrounds, training, and areas of expertise of their faculty. Because most faculty have not studied, worked professionally, or trained in the discipline of technical content, undergraduate curricula in this area tends to be underdeveloped—with some notable exceptions.

  Local needs and constraints, too, play a big role in shaping the type, orientation, and location of programs and thus the preparation that students receive for entry-level positions in different facets of the field. The nature of the program, including degree name and curricular offerings, is largely shaped by institutional location. The vast majority of technical communication-adjacent programs still reside in English departments, which means those programs share resources and budgets with literature, creative writing, composition, and other areas within English departments. Programs with few technical communication faculty also often recruit faculty from composition, creative writing, or linguistics to teach courses in their curricula. Stand-alone technical communication programs, those that report directly to a college or school, are growing but are still small in number; these programs often have more autonomy in shaping curricula, developing advisory boards and internship programs, and recruiting faculty, and they are more likely to offer a BS degree option than programs housed in English or other humanities departments.

  Institutional location impacts whether a program offers an emphasis or a full-fledged undergraduate major. Lisa Melonçon (2014) found that only 14% of technical communication or related programs offered full-fledged majors, which tend to be housed outside of English departments. Full-fledged majors offer a curriculum focused exclusively on the theories, research methods, and practices that inform the field of technical communication (commonly referred to as TPC in academic contexts); students take a required core of courses in the field and then several elective courses that help them to specialize. Programs offering full-fledged majors tend to be the most innovative and responsive to industry needs. Emphasis majors, what some scholars in the field have referred to as "emphasis degrees" (Anderson, 1978; Melonçon, 2014), offer a more divided curriculum, with core courses focused on the primary subject of study (e.g., English, communication) and elective courses focused on an area of emphasis (e.g., professional writing, media studies).

Most emphasis majors require students to take three or four elective courses in their emphasis area. Melonçon (2009, 2012) found that most emphasis majors require only a basic course in technical communication (which also serves as a service course for other departments), a capstone course, and an internship course. Citing her 2012 study, Melonçon claims that these kinds of degrees "are widely divergent and are more similar to undergraduate certificates" (2014, p. 181). These degrees "are more generalized because they have fewer courses to work with than full-fledged [technical and professional communication] degree programs" and thus cannot offer students a broad and deep foundation in the field (p. 181). The relatively few number of students graduating with full-fledged majors versus emphasis majors may indeed be a factor in the extent to which graduates are interested in and prepared for entry-level positions as technical content professionals.

Given the many constraints and challenges facing technical communication undergraduate programs, in addition to the array of career paths students are prepared to pursue in different types of programs, it is not surprising that few academic options have developed curricular pathways for students who might be interested in pursuing a career in the discipline of technical content. Further, because the content industry is a moving target, it has been difficult for researchers and faculty to stay abreast of changing roles and activities and their impact on hiring needs. Several researchers have extensively examined job ads in the field with the goal of identifying new roles, expected skills and competencies, and hiring trends (see, e.g., Brumberger & Lauer, 2015; Lanier, 2009; Stanton, 2017), but these studies focused on jobs across different components of the field's collective identity and thus did not offer specific insight into hiring practices in the discipline of content.

Changing Hiring Needs for Entry-Level Content-Focused Positions

We report in this section on the skills and qualities that hiring managers and key voices in the industry side of technical content identified as most needed for entry-level content-focused positions. We additionally report on the varied challenges that these leaders have faced when hiring for particular roles and skill sets and on the strengths and deficiencies that they have observed new college graduates bringing to their organizations and to the larger technical communication profession. We further share the job search stories of recent alumni of undergraduate programs in technical communication and related fields who are now working full time as technical communicators.

In our 17 interviews with industry leaders and in our survey of hiring managers, we asked several questions about how changing roles and activities on technical content teams are impacting hiring needs and the skills and qualities now required for entry-level positions. For our survey, we specifically recruited technical communicators in management roles who had overseen the process of hiring one or more entry-level or advanced-level technical communicators within the past five years. Our primary goal in doing this was to learn about their experiences with and perceptions of the preparation gap.

In our interviews with recent alumni of undergraduate programs who were working full-time jobs as technical communicators, we asked several questions about their experiences applying for positions in the field and about the kinds of jobs for which they felt qualified and unqualified.

Table 4.1 presents the questions related to hiring needs and job search processes that we asked of these groups. Although we did not ask questions about specific areas within the discipline of content, and we did not use it as a term (that term came to us as we analyzed the research that we conducted for this book), we did recruit industry participants who were working with or in organizations that treat content as computable data so to gather perspectives that describe hiring needs in the content discipline and that point to the skills and qualities sought for roles aligned with different areas within the discipline.

Descriptions of the Positions for which Managers Had Hired

While we did not specifically ask our interviewees from industry if they had served in a hiring manager role, many of them talked about their experiences hiring writers and content specialists for entry-level positions. Many of them, in their consultant roles, had advised hiring managers and teams on the skills, qualities, and experience needed in new hires to support content goals and initiatives.

We did specifically ask all survey respondents if they had overseen the process of hiring one or more entry-level or advanced-level technical communicators. Of the 52 respondents, 47 (90%) indicated "yes." We asked these respondents to then describe the types of positions for which they had hired and to include, for each type, the title/s, level/s, and a one- or two-sentence description of essential functions. Table 4.2 presents a breakdown of the most common position types for which survey respondents were hired between 2016 and 2021 (we conducted the survey in 2021).

TABLE 4.1 Survey and Interview Questions Focused on Hiring Needs

Survey and Interview Questions	
Survey—Managers	Of the respondents who indicated that they had overseen the process of hiring one or more entry-level or advanced-level technical communicators, we asked the following open-ended questions. • Describe the types of positions for which you have hired (for each type, include title/s, level/s, and a one- to two-sentence description of essential functions) • In the organizations in which you have worked in the past ten years, how have hiring needs changed for entry-level technical communicator jobs? • What skills and competencies does your organization look for in entry-level hires for technical communication jobs? • What strengths and deficiencies are you seeing new college graduates bring to your organization? To the technical communication profession?
Interviews—Industry Leaders	• How do you see hiring needs changing, particularly at the entry-level and particularly for content-focused jobs? • What skills and competencies do you [or do organizations now] look for in entry-level technical communication hires? • Are you seeing organizations hire college graduates from technical communication undergraduate programs? Certificate programs? • What strengths and deficiencies are you seeing new college graduates bring to the profession? How about college graduates with technical communication degrees?
Interviews—Academic Leaders	• How do you see hiring needs changing, particularly at the entry-level? • What skills and competencies are you seeing employers looking for in entry-level technical communication hires? • What kinds of technical communication jobs are your graduates getting? What accounts for that?
Interviews—Alumni	• Tell us about your experience applying for jobs in the field. • What jobs did you feel qualified for? Not qualified for? • What, if anything, surprised you?

The vast majority of entry-level and more advanced-level positions had "technical writer" in the title, with "user assistance developer" coming in a distant second. A handful of reported titles for entry-level positions included "content developer" or "information developer." Because these were positions in which writers would be working primarily with technical content, such as online help, user guides, API documentation, release notes, and UI text, we were not surprised to find that "technical writer" was the most commonly used title for new positions.

TABLE 4.2 Breakdown of Most Common Position Types for Which Survey Respondents Had Hired

Sample Titles	*Sample Position Descriptions*
Entry-Level Positions	
Junior technical writer, technical writer I, technical writing intern, user assistance developer associate, content developer, information developer	Develop and maintain user assistance for a suite of engineering software products: includes online help, reference documentation (e.g., installation and administrator documents, release notes, API docs), and UI text (writing/reviewing).
	Work with product teams and user representatives to create user guidance and installation instructions.
	Essential functions: Technical writing skills/good command over English language (oral and written), comfortable interacting with new concepts and technical ideas, highly organized with attention to detail/consistency.
	This person should hold a degree, be willing to interact with developers, be willing to handle a project alone, and should ensure timely delivery.
	Document software features, formalize procedures, and streamline authoring processes.
	Contribute to projects with supervision by staff/lead writers. Given ownership for a product area/component.
	Capable of authoring structured content in XML to a predefined document type definition (DITA, DocBook, custom markup) and with experience crafting content appropriate for reuse across multiple delivery channels.
	Work in partnership with the Product Team and Engineering Leads to project plan and develop high quality, best-in-class product documentation. Participate in editorial reviews.
	Essential functions include understanding software requirement documentation as well as oral explanations from software developers and converting them to jargon-free instruction documents for a non-technical audience.

(*Continued*)

TABLE 4.2 (Continued)

Sample Titles	*Sample Position Descriptions*
	Interface with SMEs, create and update docs in product area assigned, create outlines, report to manager with regular status updates, meet or exceed goals and deliverables expectations.
Mid-Level Positions	
Associate technical writer, professional content developer, user assistance developer	Design, write, and edit technical content for a variety of audiences, with an emphasis on usability, effectiveness, maintainability, and reusability. Proven track record of creating user guides, release notes, training materials, and FAQs.
	Continually and proactively gather information from software developers, architects, product managers, and other subject matter experts about new features and upcoming product changes. Work in an Agile environment to develop DITA-based documentation for delivery as PDF and HTML.
Information architect	Technical and organizational content architecture, DITA experts.
Advanced-Level Positions	
Senior technical writer, principal technical writer, senior user assistance developer, technical communications specialist	Create back-end content for product integration and connections to other products, as well as API descriptions. Deliverables produced have been both external customer-facing and internal reference.
	Write end-user documentation for business processes, in-app help, what's new information, and UI text. Partner both with an engineering development team and with a user assistance team.
	In addition to all requirements for junior writers, senior writers must be able to: work independently and pair with other writers as required; perform roles such as peer editing, mentoring, and managing small teams of junior authors; understand Agile methodology; must have organizations skills to be able to manage multiple projects with competing deadlines; communicate highly complex topics with clarity;

(*Continued*)

TABLE 4.2 (Continued)

Sample Titles	*Sample Position Descriptions*
	and review/test UX designs/test software to ensure usability and clarity of UI text. Well versed with DITA and single-sourcing tools.
	Independently manage contributions to multiple projects. Given ownership of multiple product areas or an entire product.
	Implement DITA component content management system; migrate Word content to DITA; edit regulatory documentation; develop publishing tools using XSLT and ANT. Manage Git repository in DevOps. Job title evolved into Regulatory Content System Architect.
	Lead projects and teams, mentor junior writers, initiate process improvements, and be always aware of new trends and technologies in the technical writing industry.
	Using independent judgment and analysis, prepare complex software-based content with minimal guidance, and organize content for rapid deployment and a high degree of customer satisfaction.

This finding aligns with results of the Write the Docs 2023 Salary Survey (Write the Docs, 2023), which asked members of the Write the Docs community to answer questions about salaries, work environments, and perspectives on job-related issues (n = 1,017). "Technical Writer" was the most widely used exact title at 32%, followed by "Senior Technical Writer" at 18.8%. The phrase "technical writer" in fact "appeared in the top five job titles and in 67.9% of all job titles." The finding also aligns somewhat with those of Eva Brumberger and Claire Lauer (2015), who found in their analysis of 914 job descriptions that 52% of titles fell in the technical writer/editor category and 23% fell in the content developer/manager category. These two categories represented jobs that largely intersected with technical content; no distinction was made between entry-level and advanced-level titles in their study.

In our analysis of the position types for which our survey respondents had hired, we did not see evidence of a decline in the use of "technical writer" in favor of more encompassing titles such as "content designer" and "information developer." In fact, none of our survey respondents listed "content designer" as a position for which they had hired. We suspect this to be the case because most respondents were managers of technical documentation teams that may have interfaced with UX teams but that served different purposes and roles in content workflows. Although technical writing activities fall under the umbrella of the disciplinary area of content design, as we argue in Chapters 2 and 3, "content designer" as a job title is still largely associated with UX research and design work as opposed to technical content work.

We certainly see the disciplinary area of content design as a good fit for technical content development activities (i.e., technical writing), which do not fit as nicely under content strategy, content operations, or content engineering. Position types within content design, however, will continue to fall into two main groups, at least for some time: those who primarily focus on in-product copy (e.g., UX writers) and those who primarily focus on non-product or "technical" copy (e.g., technical writers and user assistance developers). Required skill sets for these two groups overlap in some ways (i.e., strong audience analysis, research, and collaboration skills) but differ in other, substantial ways, as our findings reveal.

Several themes emerged in respondents' brief descriptions of the essential functions of the entry-level technical communicator positions for which they had hired. Essential functions most mentioned included the following:

- **Work closely with product teams.** Writers are expected to interact with SMEs such as engineers and developers, often in an Agile environment, to help plan projects and develop service and support documentation. Ability to understand product requirements and SME explanations, as well as familiarity with product development processes, is a must. So too is the ability to ask good questions to get the right information and to give and apply feedback.
- **Possess strong technical writing skills.** Writers are expected to develop and maintain documentation (e.g., online help, reference materials, installation instructions, API docs) for one or more products and need not only a strong command of the English language but also the ability to author structured content in accordance with specified content standards (usually, but not always, with goals of reuse and multichannel delivery in mind). Writers hired to work in software development contexts often work on Agile project teams and, in some contexts, are expected to document code using a docs-as-code approach and manage documentation in source control systems, most particularly GitHub.
- **Handle projects independently.** Writers are expected to own one or more products or product areas and manage timely delivery of product documentation with some supervision. Because onboarding resources tend to be scarce, writers need to get up to speed quickly and be ready to move forward on assigned projects with minimal hand-holding.

In addition to these functions, respondents also often mentioned that the entry-level positions for which they had hired required one to three years of experience, a requirement given the complexity and fast pace of the environments in which writers were expected to work.

Some respondents offered titles and brief descriptions of essential functions for the mid-level and advanced positions for which they had hired (positions within the area of content design). Titles for mid-level positions included prefixes such as "Associate" or "Professional," whereas titles for advanced positions included prefixes such as "Senior" and "Principal." Notably, the most common titles for these positions were the same as those for entry-level positions—technical writer and user assistance developer. Mid-level positions tended to require four to seven years of experience with advanced positions requiring eight years or more. A key difference we saw between entry-level and mid-level descriptions was the expectation that writers would contribute to multiple projects and own multiple products or product areas. Advanced positions were more strategic and supervisory, with candidates in these positions being expected to define vision, drive initiatives, and lead multiple products and teams. Experience working in complex software environments and overseeing system implementations and content migrations was a common expectation for advanced positions.

Figure 4.1 shows how responsibilities increased from entry- to mid- to advanced-level user assistance developer positions in a Fortune 500 software company, as reported by a Senior Director of User Assistance who had overseen hiring processes for the different levels. As reported by this informant, the user assistance team primarily created structured content, managed it in both a component content management system and source control system, shared it across multiple teams, and published it in a variety of formats. The team also created videos and animations, as well as user interface text.

Entry-level user assistance developers at the Fortune 500 software company were tasked with developing a full set of content deliverables for one product or product area, and they received mentoring and guidance from more senior members of the team. In her survey response, the Senior Director of User Assistance wrote that entry-level positions at their company required a technical background and experience with structured authoring—candidates with computer science degrees were preferred.

Another informant, a Director of Technical Documentation for a Fortune 500 enterprise hardware company who had worked in management for 25+ years, described five levels of information developers that she had hired in the previous ten years. The technical documentation team primarily created structured content, managed it in both a component content management system and source control system, and published it in a variety of formats. The team also created user interface text.

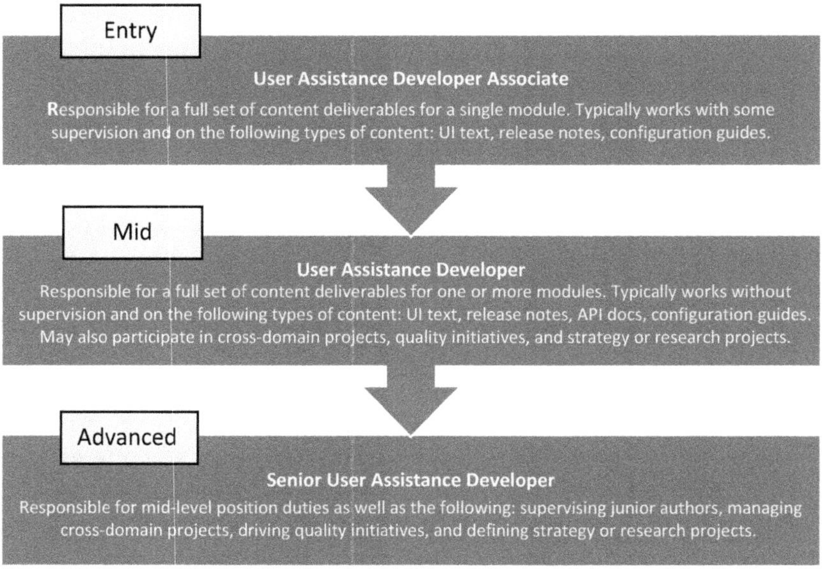

FIGURE 4.1 How Responsibilities Increased from Entry- to Mid- to Advanced-level User Assistance Developer Positions in a Fortune 500 Software Company.

Figure 4.2 shows how responsibilities increased through the levels. For entry-level positions, computer science majors or minors were preferred. Information Developer 1 positions did not require experience, opening the door for college graduates to apply. The company had a structure in place that both encouraged and expected advancement through the levels. Because of the highly technical environment into which information developers were hired and given the Director's experience in hiring both computer science graduates and technical communication graduates, the Director reported that she had learned that she could train computer science graduates for the position much quicker than she could train technical communication graduates. As to why, she remarked: "If I hire CS majors, we have to teach them to organize info. If I hire TW majors, they don't seem to know enough programming/data concepts to swim fast." As we discuss later in this chapter, a preference for hiring computer science majors for entry-level technical content positions was evident throughout our interview and survey findings, and we heard a similar preference in our discussions with podcast hosts Scott Abel and Patrick Bosek (Abel & Bosek, 2022) and with participants at the 2022 CIDM Best Practices Conference.

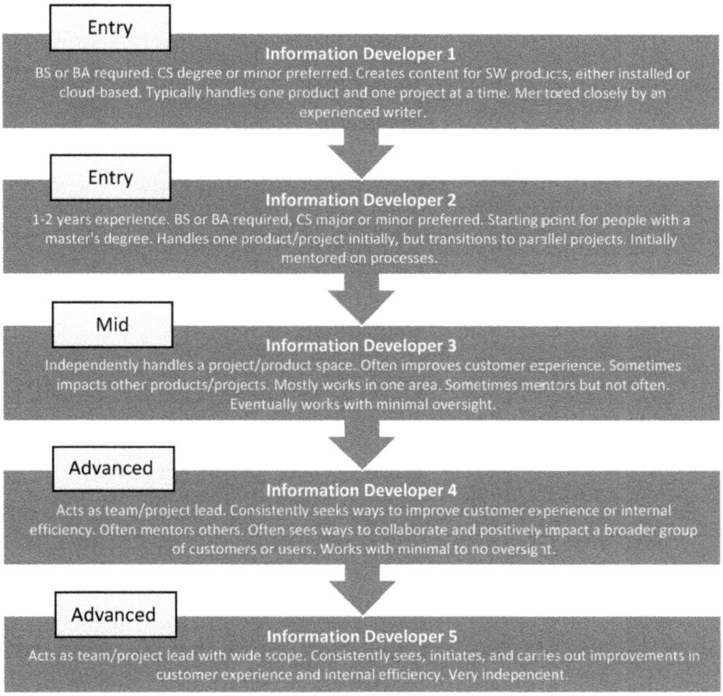

FIGURE 4.2 How Responsibilities Increased Over Five Levels of Information Developer Positions at a Fortune 500 Enterprise Hardware Company.

Skills and Qualities Most Needed

In this section, we present themes on the question of how hiring needs are changing for entry-level content-focused positions—as reported by managers, consultants, recruiters, and academics, across survey responses, interviews, and focus groups. A majority of the managers who participated in our research led teams that worked primarily with structured, technical content in XML or Markdown environments. Although we did not ask survey respondents to indicate which industry they represented, responses suggested that most respondents worked in the technology, information technology services and solutions, and manufacturing industries, with documentation of hardware and software being a key job responsibility. Themes summarized in this section are specific to hiring needs and challenges in the technical content space, which is increasingly being shaped by the forces of automation, artificial intelligence (AI), and complex systems coupled with the need for organizations to publish content to many channels, provide end users seamless content experiences, and engage in

FIGURE 4.3 Reported Skills and Qualities Most Needed for Content-Focused Jobs.

continuous content development as Agile methods and approaches become the norm.

Figure 4.3 represents the four broad skills and qualities needed for content-focused jobs most mentioned by our research participants, the most prominent being technical knowledge and skills, followed by having a learning mindset—that is, being curious and inquisitive and unafraid to learn new tools and processes. Notably, whereas this section and Figure 4.3 focus on reported skills and qualities needed for content-focused jobs, Chapter 6 focuses on what our research participants had to say about implications of changing roles, activities, and hiring needs for educational programs, particularly undergraduate degree programs, that aim to prepare future content professionals. In response to questions about educating future content professionals, participants expanded on the skills and qualities reported here, and based on those responses as well as existing literacy frameworks in the field, we introduce in Chapter 6 a competency framework for content professionals.

At the time that we conducted our interviews and survey, generative AI tools were in their infancy and had yet to gain wide-scale public attention. We discuss here the hiring needs and challenges reported before the large-scale adoption of ChatGPT, Writer, and other generative AI tools in early 2023.

Technical Knowledge and Skills

All of the managers, recruiters, and consultants who we talked to, as well as many of the managers who responded to our survey, emphasized the need for entry-level writers who have a love of technology, a willingness to play and problem solve with tools, and an ability to pick up complex technical information quickly. Further, a good number emphasized the need for entry-level writers who have a background in computer science and expertise in a technical domain area. Technical fluency expectations intersected with the need for hires who are comfortable working in structured authoring workflows and in Agile workflows, which require continuous content development, close collaboration with developers, and an ability to read some level of programming code.

Because so many of our research participants spoke to the growing importance of technical knowledge and skills for entry-level content-focused positions, we identified key themes that came up over and over in discussions of hiring needs and then teased out reasons accounting for why particular qualities or knowledge and skills sets were so highly sought out. Key themes were as follows:

- Ability to grasp technology fast
- Eagerness to use technology to problem solve
- Proficiency with computer science concepts and programming
- Expertise in a technical domain.

Ability to grasp technology fast. New hires must be able to grasp technology fast because products are becoming more complex; updates and releases are continuous; writers are expected to own a product area and begin contributing content immediately; different interorganizational teams often use different content authoring and management tools; and resources for training and mentoring have decreased significantly across industries.

The following testimonials from our interviewees and survey respondents point to reasons why managers are looking for hires who can grasp technology fast.

- **Complexity of products.**

  Entry-level people need to embrace technology FAST! The role involves learning, but learning on a broad range. These people need to attend product training, and they need to come up to speed on tools. Tooling is less shrink-wrapped. We use Git, XSLT, Oxygen XML, and Jekyll. Oxygen is the only shrink-wrapped tool, and it's just half-wrapped. The product is complex and interfaces with a broad range of other technologies. We have different deployment scenarios that use different technologies.

—*Survey respondent*

...all of the products have become modular and variant rich. So each machine looks different than the other one because people can just use a product configuration engine and configure their product, and that's delivered as they ordered it. And that also means you need to understand all the complexity of product variants, variant management, modularization, and metadata. This is something you need to understand because interfacing systems and aggregating data from a different system become more and more important. That means as a technical communicator you work with very different IT systems and maybe you also create the hub that connects all those systems. You need to know how to manage IT projects, work in an agile way, define requirements, organize sprints, and look at the results of sprints and stuff like that.

—*Founder and CEO of content solutions consultancy (interviewee)*

- **Agile software development.**

Moving to Agile software development has also put pressure on faster ramp-up time. There's far less grace period for new writers to get up to speed on tools and concepts. New writers who struggle to engage confidently and proactively with their SMEs will struggle to find success in a faster-paced environment. They often need to be able to jump into using the docs-as-code approach to documentation.

—*Survey respondent*

In recent times, product development time has reduced drastically. There are no longer any long releases for software products in most companies. Entry-level authors are expected to work with multiple stakeholders within documentation teams as well as with other stakeholders to complete their deliverables on time. This requires them to grasp things at a faster pace and at the same time, collaborate, plan, and manage their work in a fast-paced work environment. In addition, both new and experienced authors are expected to contribute more than just delivering documentation. They will be greatly respected if they can test products to identify defects, and suggest improvements to improve quality of products, workflows, and processes.

—*Survey respondent*

- **Automation of repeatable tasks.**

> Due to improvements in technology, there are few well-defined, repeatable tasks remaining. Previously, an entry-level TC might have a year or two to get comfortable in the organization by doing document prep work (proofreading, formatting, link checking). As these things are largely automated now, the entry-level TC will get started immediately on contributing content. To start, they will be making small, routine updates, but then they will progress usually in about three months to owning a domain.
>
> —*Survey respondent*

For entry-level content-focused positions, automation of repeatable tasks has meant new writers need to be able to get up to speed quickly to start contributing technical content. In most contexts, contributing technical content means authoring structured, semantic content through XML, or Markdown syntax (usually in docs-as-code workflows). Alex Masycheff (2018), in "Keeping Technical Writing Jobs Relevant with Content Automation Technologies," suggested that the increased automation of "low value tasks" would result in technical writers needing to "deal with more complicated and creative tasks" and seeing their job as interdisciplinary, including writing "but also information design, UX, customer support, user engagement, and other areas humans are traditionally good at." Writers who have the technical knowledge and skills required to get in the door, then, should have more opportunities to contribute to strategic efforts in meaningful ways.

Eagerness to use technology to problem solve. In addition to the need to grasp technology fast, entry-level technical writers need to demonstrate a love for technology and eagerness to learn how it works; familiarity with structured authoring tools and content management workflows, including source control systems; and a desire to use tools to improve processes and outcomes (e.g., content quality, user experience, workflow governance). In our conversations, consultants in particular emphasized the importance of curiosity—of wanting to understand how tools and technology work.

The following testimonials point to reasons why hiring managers are looking for candidates who demonstrate a love for technology, familiarity with commonly used tools, and an eagerness to use technology to problem solve.

> I find it very difficult to work with technical writers who are afraid to touch the product, afraid that they're going to break the product and the like.... So an appreciation, if not a love of technology is certainly

> one aspect that I look for. And it doesn't matter if you're writing for software development or you're writing for a bank. That love of technology, it's more to how are you going to make your job easier? How are you going to make delivery and consumption of that information better? By embracing technology.
>
> —*Principal consultant and co-founder of an authoring solutions company (interviewee)*

> So… some degree of facility with the tools, sort of the technical side of technical communication, is obligatory, I think. They have to feel comfortable with being able to work with content that may, at some point, become XML. Maybe DITA. And…working with, as I was just wrestling with moments ago, the DITA Open Toolkit. And trying to get it to do something that it just doesn't seem like it wants to do today. And bashing about until you get an output, and that can sort of support an experience that you want.
>
> —*Principal researcher and managing director of a content research consultancy (interviewee)*

> And I do think that there's an expectation at this point that they have tool familiarity, and we're not talking just Microsoft Word, but we're talking about whatever tool. Companies expect that you've got some exposure to structured authoring, whether that's in FrameMaker or DITA, or whatever, but some tool familiarity, which of course is, to me, the easiest thing to train, once you get there, but there's an expectation that that is already in place with the people that are being hired. They expect to train you on the product that you'll have to document, but they expect you to still come out of school with the ability to use whatever the right tools are and create content from that, well-written content, not having a lot of errors of copy-editing or that type of stuff as well.
>
> —*President and owner of a content solutions consultancy (interviewee)*

While in the past familiarity with tools such as FrameMaker was a plus, experience with tools and standards for authoring structured content is now a requirement for most entry-level content-focused positions, in the same way that familiarity with Microsoft Word and desktop publishing tools is still a requirement for authoring unstructured content. Some managers go so far as to expect familiarity with a suite of authoring, management, and delivery tools (not a realistic expectation, but one that exists). For example, one survey respondent said that, for entry-level positions, he looked for familiarity and experience with "GitHub, GitLab, AsciiDoc, Jekyll or a similar page builder, Markdown and other 'docs-as-code' systems and tools." The bottom line is that students preparing for careers as technical content professionals must truly enjoy learning and writing with technology and finding ways to leverage technology to improve processes and outcomes.

Proficiency with computer science concepts and programming. Perhaps our most surprising finding was the extent to which hiring managers said that they preferred to hire computer science graduates over technical communication graduates for entry-level content-focused positions. Although computer science graduates tended to lack strong writing and communication skills, managers overall suggested that they found it easier to teach writing and the concepts of structure on the job than to teach computer science concepts and programming on the job.

The general consensus among hiring managers who participated in our interviews and survey seemed to be that ideal candidates for today's entry-level content-focused positions were those who have a computer science background and training in both structured authoring and basic principles of technical writing. But these candidates are rare and, according to one manager with years of hiring experience, are hard to keep given their highly sought out skill set:

> Another reality is the frustration hiring managers and teams have when they lose an early talent to the competition after spending time and effort training them. So we seek but rarely find entry-level candidates that already know DITA and the basic concepts of working in a CCMS, Jira, GitHub, etc.
>
> —*Survey respondent*

Many hiring managers expressed frustration with not being able to find strong candidates who have both the technical background and strong structured authoring and other "soft" skills required for entry-level positions. The need for candidates who are well versed in how to effectively collaborate, manage projects, conduct research, organize information, and write clear and concise prose hasn't changed; the need for candidates who can do all that plus understand data structures, markup languages such as Markdown and XML, content management systems, and more has.

The following testimonials speak to the need for this more robust skill set but also to the kinds of frustrations and dilemmas that hiring managers face when reviewing applications for entry-level positions. These frustrations and dilemmas often reflect a general feeling among industry leaders that college graduates are not interested in pursuing careers in the profession of technical content.

> If I hire CS majors, we have to teach them to organize information. If I hire TW majors, they don't seem to know enough programming/data concepts to swim fast.
>
> —*Survey respondent*

> We still need entry-level communicators, but I'd say it's harder to find them. We review lots of resumes of new computer science majors who are interested in technical writing. They generally have the technical knowledge we're looking for, but often they have no writing experience, or their writing samples are just not good enough. We also look at the resumes of new grads who have stronger writing credentials (for example, a journalism or tech comm degree). But they have to show some interest in or experience with software documentation.
>
> —*Survey respondent*

> Due to the complexity of the products (SW platforms), I expect much more industry knowledge. I really need people who understand modern SW structures like containers and programming. We can teach tools. My lead writers would like to see people with training on how to organize content, but I haven't found people who know both CS and tech writing.
>
> —*Survey respondent*

The reasons for this managerial need for candidates with computer science and strong technical backgrounds vary, but primary factors include the need to document APIs, understand complex products, read code, write for an audience of developers, write alongside developers using docs-as-code approaches, and understand software development processes. The following testimonials speak to some of these reasons:

> In general, entry-level writers are expected to have more technical knowledge than in years past. It's hard to find candidates who have a good mix of technical knowledge who can also write clearly. More and more, it seems that developers expect writers to read code to understand changes to a product. In the past, writers started a project with a detailed functional specification. Now that's becoming less common.
>
> —*Survey respondent*

> What is software, how is that built? That's something that I think everyone should have an understanding of, including taking a couple of programming classes. Not because you're going to be a programmer, but understanding that end-to-end process of how software, how a program is eventually in the hands of a consumer. If I just think of… Think of any application that you use, even a browser. I'm on a computer, but how does

> that browser get to me? How does somebody decide how it's… What was the concept? What was this idea? How do you go from idea to delivery to a consumer, and how does the consumer consume it? That process, from conception to delivery, to even obsolescence, I think is something that everyone should understand. That's something that's definitely missing sometimes in entry-level writers. The core communication skills could be strong, and they should be, but it's much more than communication.
>
> —*Senior director of user assistance (interviewee)*

The need to document APIs often resides in a communication structure that has SMEs taking the roles of authors and readers. In such a structure, technical writers are expected to understand the product with the same depth as engineers. At the same time, technical writers are increasingly writing more for an audience of developers than end users. As Tom Johnson has argued in a series of blog posts on technical communication trends, the field "has undergone a major evolution from end-users to developers as the target audience," requiring writers to understand programming languages and the inner-workings of APIs (2018a); further, "as developer tools and languages becom[e] increasingly specialized and complex," developers are taking on "more active, collaborative roles in documentation" in software development environments (2018b), thus necessitating, as one would expect, writers who are well versed in computer science and can productively collaborate with developers. Writers are also increasingly expected to use a docs-as-code approach to content development, management, and deployment; this approach requires a collaborative version control system such as GitHub and a content authoring toolstack that resembles contemporary programming languages.

Expertise in a technical domain area. Though only a few interviewees mentioned the need for writers to have expertise in a technical domain area, we thought it an important enough point to emphasize here. Reflecting on his experience having worked for several organizations with complex products and highly technical content, one manager said that the product management and development teams that he has overseen have insisted on hiring entry-level technical writers with technical qualifications because they must be able to pick up complex technical information quickly. The hires have had either an engineering or computer science background.

Another interviewee, president of an employment agency specializing in technical writers and other content professionals (e.g., content strategies, content engineers, instructional designers), shared that the vast majority of candidates who he has matched with innovative organizations did not have degrees in technical communication. Rather, given the need for domain knowledge in particular industries and product management and development contexts, the most successful candidates have had degrees in

STEM fields. His following comment speaks to the value of having domain knowledge:

> Biotech companies want people with a biotech background. Networking companies want people with a computer background, software people want people with a background in software engineering. So, getting a degree in a domain gives you domain knowledge, so you can then understand what you're writing about. And, I hate to say this, not once, not once has any company ever come to me and said, "I want someone with a degree in technical writing."

Domain knowledge, however, was not widely cited as a hiring need by a majority of our research participants; the ability to grasp technology fast and an eagerness to use technology to problem solve were mentioned most frequently.

Academic perspectives. Among the 14 academic leaders whom we interviewed, there was general agreement that hiring managers wanted to see evidence of proficiency with the concepts and processes behind commonly used authoring and publishing tools. Several mentioned that students did not necessarily need to know how to use specific tools, despite job ads that included long lists of specific tools, but they did need to be familiar with the types of tools used, what they do, and how they work. For example, a student who used Oxygen Editor to create content for a class project should be able to transfer their knowledge of the tool to a different XML-based authoring toolstack. A professor and chair of a technical communication department had the following to say:

> Employers are very fond of sticking the names of tools in their ads. We coach our students, "As long as you know a tool like that one, you're fine." You don't have to know every tool that every organization uses, but you do need to know some kind of word processing tool. You need to know some project management tools. You need to know something about delivering information. You need to know some of those tools, whether it's HTML or XML or more specifically Flare or whatever it happens to be.

Familiarity with tools, at least as reported by the administrators and instructors whom we interviewed (all of whom were in programs that continuously updated curricula to align with industry practices), was an essential competency for entry-level content-focused positions. These programs had developed courses such as Intelligent Publishing and Information Architecture to ensure that students had opportunities to learn the concepts and processes behind different tool types so as to develop transferable knowledge and skills—and to build portfolios that showcased their capabilities.

When asked, "What skills and competencies are you seeing employers looking for in entry-level technical communication hires?," only one of our academic interviewees mentioned computer programming and none mentioned software development knowledge or ability to understand complex systems and processes. Half of them did mention proficiency with XML and DITA, a need that a majority of industry experts identified. As products and content operations grow in complexity, and as writers work more closely with engineers and developers co-constructing knowledge and content with input of AI agents, we expect the need for advanced technical fluency to become an even more pressing hiring need.

Learning Mindset

A second prominent hiring needs theme that we identified across interview transcripts and survey responses was the need for hires who embraced technology and who were excited to learn new tools, products, and processes. We identified this as a separate theme from technical knowledge and skills because it encapsulates a growth mindset, a concept introduced by psychologist Carol Dweck (2007) in her book, *Mindset: The New Psychology of Success*. Someone who has a growth mindset understands that they can change their abilities and that change involves failure, adjustments, and persistence. By contrast, someone who has a fixed mindset thinks that they cannot change their ability to learn and succeed—that their knowledge and capabilities are fixed and that mistakes are signs of failure.

Succeeding as a content professional in organizations that treat content as a computable asset requires taking on challenges and learning from them. Continuous technological changes and continuous product development cycles mean technical content professionals must be curious and adaptable and be willing to tinker with technology, experiment with new ways of doing things, take initiative, and ask for guidance when needed. In application materials and job interviews, hiring managers are looking to see how candidates think and approach tasks, and they want to see evidence that candidates are equipped to "jump in and swim." When presented with a messy task, they want to see an independent thinker who quickly gets to the work of problem-solving and leveraging available resources toward a solution.

A senior director of user assistance for a major software solutions company who has hired dozens of entry-level content professionals shared that she always presents interviewees with a particular scenario and asks them what they are going to do. She explained her approach as follows:

> I come to you, and I give you the name of a product and the name of one person, and I say, "Okay, what are you going to do? We have to document this. How are you going to get the documentation and what are you going to do? How are you going to approach this?" That always is very revealing, and what I'm looking for is, "What are you going to do on your own? How much are you going to drive this? How much are you going to set expectations with the team? Are you going to start with contacting that person and asking them to drive everything? When are you going to start thinking about dates?" In the corporate world everything is driven by dates, dates and deadlines.

What this senior director of user assistance wants to see is how well the candidate will be able to problem solve and handle documentation projects on their own. The ability to learn quickly and relatively independently came up numerous times in response to our question concerning hiring needs and what managers are looking for in new hires. Resources for training have declined across organizations, and mid- and advanced-level content professionals have limited time for guiding new hires through every step of a process or task. The climate for new hires, then, is one where they are expected to "come up to speed with minimal help" and "have a get up and go attitude." Being a self-starter and problem solver and having a learning mindset are critical qualities for success in this climate.

When we asked industry leaders how hiring needs have changed for entry-level positions, the number one response that we received was the expectation that new hires would learn quickly and would embrace the challenge of always learning. The following responses from hiring managers speak to this expectation.

> It has changed in such a way where entry-level technical communicators are expected to upskill quickly, learn tools quickly and start delivering documents. Also, they are expected to share their fresh ideas and thoughts on documentation delivery and other processes.
>
> —*Survey respondent*

> For entry level positions, the recognition that you aren't done learning two to three years after you've joined a company is a requisite. The drivers behind the changes are the regulatory requirements that

> constantly evolve. They are so complex that one needs to work in structured content.
>
> —*Survey respondent*

> It's people who aren't afraid to learn something new, a new tool, or a new standard, or sort of jump in knowing, a bit like transitioning between these applications to say, this is different, but I'm going to bash around until I feel comfortable.
>
> —*Principal researcher and managing director of a content research consultancy (interviewee)*

> Broadly, we're looking for people that are smart, and that like to learn, and think learning is fun because the thing that really distinguishes what we do specifically from other jobs is that it's going to be different every week, every month, every year. There's always going to be a new tool set, there's always going to be a new requirement, a new something. So, we're looking for people that think that's fun as opposed to the people who want to go do the same thing for the rest of their career.
>
> —*CEO and founder of a content solutions consultancy (interviewee)*

Three academic leaders from our participant pool also touched on the theme of having a learning mindset in their responses to our question about what they have noticed about changing hiring needs for entry-level technical communication positions. These leaders emphasized the need for new hires to embrace technology and not be afraid to learn new tools and methods for communicating through new tools. One interviewee had observed an expectation among smaller employers that recent college graduates, given their youth, were prepared to work with a wide range of technologies and could draw on their extensive technical know-how to meet the employers' expressed needs. Many employers assume college graduates are enthusiastic about technology and eager to learn new tools. This is certainly the case for most students majoring in technical communication, and it presents a real opportunity for programs to cultivate this enthusiasm and curiosity.

Structured Authoring and Content Standards

The ability to create topic-based, well-structured content was widely considered a foundational skill among hiring managers for content-focused jobs. Candidates were not expected to know particular tools or standards, but they were expected to have a conceptual understanding of structured authoring as a methodology and to be familiar with the technologies and standards available for supporting structured authoring.

This hiring need aligns with Keith Schengili-Roberts' (2022a) statistical analyses of job postings for "Technical Writer" on Indeed.com between August 2012 and August 2022. Schengili-Roberts identified a significant spike in 2021 in the number of job postings seeking writers with XML experience and familiarity with one or more content standards such as DITA. In 2022, Schengili-Roberts (2022b) also saw an increase in job postings mentioning Markdown and GitHub, suggesting that more companies were adopting docs-as-code approaches to structured authoring. In the 2018 issue of *Intercom* that shared results of the state of the technical communication industry census, Scott Abel (2018) reported that the creation of structured content was a growing trend on teams around the globe, with 51% of teams surveyed reporting use of XML to create content (55% of those teams reported using the DITA XML schema) (p. 37). In that same issue, Saul Carliner and Yuan Chen (2018b) reported on the work practices that most affect technical communicators—standards and structured writing played the biggest role in work projects (p. 15).

Because of the wide adoption of structured authoring across the content industry, hiring managers expect college graduates of technical communication programs to have experience creating structured content. The following testimonials from managers and consultants speak to the need for entry-level writers to have experience creating structured content and working with content standards such as DITA—which we discussed in Chapter 3.

> So, skills like clear writing, structured authoring, information typing, chunking content appropriately. Those are all foundational skills that … apply very nicely when one takes technical content that was produced today or even in past years and uses that content as the basis for a chatbot or a voice application. So, there's a combination of foundational skills. Some skills have been with us for some time, like XML authoring as an expression of the structured authoring. Metadata happens to become more and more important when you're delivering content in different ways, but those aren't new skills. Those are just skills that technical communicators have been using for some time that are now even more valuable than they had been.
>
> —*Content services solutions engineer (interviewee)*

> Entry-level writers are expected to have an understanding of the DITA methodology, even if the authoring tools used differ from organization to organization.
>
> —*Survey respondent*

> We want entry-level writers to have some DITA experience or at least understand the fundamentals of structured writing.
>
> —*Survey respondent*

> When they're [hiring managers] looking for new hires, they want the people who already know how to do that [topic-based writing]. Even with right out of school, they're hoping that that's the kind of education people are coming in with, that they can help guide as well, and pick up the ball easily.
>
> —*President and owner of an information development and management consultancy (interviewee)*

A senior director of content strategy who has worked in high-level management for multiple companies and has hired dozens of writers in her roles overseeing content teams offered the following insight into what she looks for when hiring. We asked her, "What are the skills and competencies that you're looking for when you're hiring somebody who works with technical content? What are you looking for that you don't find out there?"

> For me specifically, and this is specific to how I operate my team, I look for structured authoring. I'm not overly concerned if they know DITA or DocBook, but I want them to understand why DITA, why DocBook, like pull back the covers and understand what the overall benefit of some structure is and really be able to answer some detailed questions about that.
>
> I might look for tools on resumes, not because I care which structured tools they've used, but because I know if they've used certain tools, then they abide by the rules of DITA and if they use other tools, then it's a squishy DITA. And so I know the tools well enough to know that if they come from MadCap Flare, they're not going to have a good understanding of structured content. They're going to think they do, but they won't.
>
> I have hired people right out of college. When you have a program that teaches structured authoring, in some ways, bringing someone straight out of school is as good, if not better, than bringing someone who has their mind set on a certain thing. And so I definitely think there's a place to hire directly out of college. It would just require a very specific skill set.

For this hiring manager and several of our interviewees, structured authoring is a way of thinking. It's a way of understanding the potential of content to become useful information, accessible whenever and wherever a user needs it. When interviewing for entry-level positions, hiring managers want to see how applicants think about structured authoring and how, given different documentation scenarios, they would employ a structured authoring approach. For example, one director of a content research consultancy, who regularly hires for content-focused positions, looks to see what kinds of questions applicants talk through when describing how they would approach a scenario. This director wants to see questions such as "What is it we really need to say? What is it that people really need to know to do or not to do? What's the prioritization, and how does this documentation project fit into the overall product context?"

While proficiency with structured authoring is a highly sought-out skill set for entry-level content-focused positions, it is not a skill set that hiring managers often see among recent college graduates of technical communication (and related fields) programs, as is evident in the following testimonials:

> I think they're still writing documents. They're not writing content. They still don't have a good sense of structure. We do still see problems with grammar and usage, but that's relatively minor compared to the fact that I'm looking at a procedure, and the steps are in the wrong order. Or I see a number, and it's the result of a step, not a step. It's like, okay. They don't get it. They'll be rejected, just like that.
>
> —*Information architect and CEO of a communications consultancy (interviewee)*

> The things we're looking for are… We like to see XML. We like to see DITA. Those are pretty rare. If we don't see those, we're looking for some sort of markup language, usually HTML.
>
> —*CEO and founder of a content solutions consultancy (interviewee)*

> We seek but rarely find entry-level candidates that already know DITA and the basic concepts of working in a CCMS, Jira, GitHub, etc.
>
> —*Survey respondent*

> The only way we are going to be prepared for the next, as Karen McGrane says, "Content Zombie apocalypse," is to divorce structure from output. Because how are you going to be prepared for the next toast printer (sic) that comes along. If your format is not separated, if content is not separated from format. So, if a company is doing structured authoring, they want a person to have had that concept already. And I don't think a lot of certificate programs are teaching that.
>
> —*President of a content professionals employment agency*

It is true that a majority of undergraduate programs in technical communication have yet to integrate structured authoring into core curricula, and even fewer have created curricular pathways for students wanting to pursue careers as technical content professionals. But we have seen wide recognition among our academic peers, including those whom we interviewed, of the need to embrace structured authoring as a core skill set of today's technical communicator.

Strong Communication and Writing Skills

As would be expected, strong communication and writing skills are highly valued for entry-level positions, but they are no longer the most valued. Advanced technical fluency and proficiency with structured authoring are the skills that managers and consultants say are most needed for success working with technical content in organizations that have adopted digital content transformation workflows.

When having to decide between someone with advanced technical literacy and someone with strong writing skills, most managers and consultants in our study said they go with advanced technical literacy (i.e., with someone who has a computer science or engineering background). When the choice was between advanced technical literacy and structured authoring proficiency, the line was less clear. But what we heard over and over was that, by and large, recent college graduates rarely have both advanced technical literacy and proficiency with structured authoring; rather, they tend to have either advanced technical literacy or strong unstructured writing skills. So while the ability to write clear and concise prose is highly valued, as is the ability to write effective reports, proposals, and other professional genres, these writing skills were mentioned less often than the ability to write effective topic-based content in structured authoring workflows.

Managers and consultants also talked extensively about the importance of strong communication skills for entry-level positions, particularly the "soft skills" of conducting research, interviewing SMEs, and listening. Problem-solving and process thinking skills were also emphasized.

> It has become more competitive now. We no longer look for only written and verbal skills. Along with these, we also try to touch upon their social skills, behavior, problem-solving skills, leadership skills, and so on.
>
> —*Survey respondent*

> Writing skills are not enough—technical and process skills are important. While decent writing is important, critical thinking and process thinking are more so. Writing is easier to teach than thinking.
>
> —*Survey respondent*

> One of the things I always ask for when I interview is writing samples. I'm not going to read through whatever someone provides. First I want to see what they consider their pride and joy, that represents their role as a technical communicator in a company. Fresh out of college you'll bring something that you did for a class or maybe at an internship, you've prepared something. I want to see what you bring and then I want to talk about it. How did you arrive at that? What was the process? How do you approach writing in general? Do you start outlining, do you start playing with the product? I really want to see how you arrived at the end, at the end product. That's important to me. It's how methodical you are in your thinking, in your approach. That's something I definitely look for.
>
> —*Senior director of user assistance (interviewee)*

> We're looking for things like interviewing skills. Like, can people have the soft skills to talk to other people and have a reasonable conversation? Because quite often the answer is no.
>
> —*CEO and founder of a content solutions consultancy (interviewee)*

> The ability to listen and absorb. If you're talking, you're not listening. You should be able to go, "What does this do?" And then shut up. Take it in, and then use that. So, you have to listen and absorb that knowledge.
>
> —*President of a content professionals employment agency (interviewee)*

Strong writing and communication skills, including research and interviewing skills, have always been key learning outcomes in technical communication curricula. So too has been the ability to tackle communication problems through rhetorical analysis, project management, tool use, and application of writing and communication skills. Emphasis on metacognition, too, benefits students in that it teaches them how to reflect on their process and rhetorical decision-making and to justify their content, design, organization, and writing style choices. Our research suggests that these skills are still important, but students wanting to pursue careers in the technical content facet of the field also need to develop technical knowledge and skills, a learning mindset, and proficiency with structured authoring.

Hiring Challenges for Entry-Level Positions

In the December 2021 issue of the CIDM Best Practices newsletter, Dawn Stevens and Kathy Madison (2021) reported on hiring and training practices in the CIDM community, which is composed largely of information-development, training, and support managers. They conducted a member survey to find out what qualifications were most important when making hiring decisions, how easy or difficult it was to find qualified candidates, and what strategies members used to recruit qualified candidates. Stevens and Madison found that "less than 25% of respondents even entertain hiring people directly out of college, preferring to hire someone with experience." And only 19% of respondents felt that recent college graduates were well prepared for the job (p. 13). Respondents were not seeing recent graduates who had enough experience and know-how to do the job without extensive training and mentoring, and their organizations did not have the resources to support comprehensive onboarding programs.

This finding reflects some challenges that managers in our study had faced when hiring for technical content-focused positions. The biggest challenge was that of their organization's decreased investment in inexperienced hires. Technical content teams across the board had fewer resources and less time to dedicate to training new hires, which meant that new hires needed to be able to get up to speed quickly and start contributing content quickly. Several survey respondents reported that they did not hire new college graduates because they did not have time to train them. A senior technical publications manager with 30+ years of experience hiring for a wide range of technical content-focused positions for different companies spoke about budget restrictions and what he sees as an investment decline across the industry in onboarding programs.

> Because we've hired so many fewer entry-level or even grad-school people [referring to the content industry], my concern is we've almost lost the ability to develop effective onboarding for them. If we haven't, and I haven't recently, hired a grad because there's no budget for it, they're jumping in and having to sink or swim where everybody assumes there's a certain level of acumen. That whole desire to mentor them and bring them along and test them and shelter them a little bit, to be honest with you, so that they can grow—we've lost those understanding programs. We'd have to ramp them up, we'd have to regenerate them, which we can do, but they're not out there.
>
> I always have a choice. You can get a principal or you can get an entry-level writer, but one way or another, you're responsible for getting this thing out the door. I would say more than nine out of ten managers are just totally risk-averse these days. They're just going to go with the principal.

Changing hiring needs for technical content-focused positions, coupled with reduced training resources, has meant fewer available entry-level positions in the discipline of content, and those positions that are available tend to require a level of technical skills that many graduates of technical communication programs do not have—the result being that graduates either don't apply or do apply but feel unqualified (we document some job search experiences of recent alumni in the next section). Given the technical skills required, entry-level and mid-level positions are often filled by people pursuing technical communication as a second career; these professionals tend to come from technical fields such as engineering or software development and thus already have the strong technical skills required for some content-focused positions. They might require less mentoring and onboarding than new college graduates.

Hiring needs and willingness to invest in recent college graduates, however, are different across contexts and industries, and some companies understand the need to grow talent. One CEO of a single-sourcing solutions consultancy who regularly works with content teams in the Bay Area explained what she had observed:

> In the valley, they expect you to jump in and swim. There is no training, there is no time to tell you every day what every step you need to do is. It's just not that way here.
>
> [But] I will say that even here in the valley, there are some companies that are more inclined to hire new graduates, because the people who are in management positions understand the need to grow people. There are a couple of companies here that will take someone and let them grow. But that is a cultural thing inside the company and the people doing the hiring. If you have a manager who's not inclined that way, they [graduates] are never going to get in there.

Despite fewer companies investing in mentoring and onboarding for content-focused positions, there is a general consensus among industry leaders and managers of content teams that a new generation of technical content professionals is needed. Managers want to invest in recent college graduates, particularly in contexts where senior content professionals are planning to retire.

Required and desired qualifications for entry-level positions, however, have changed, and managers are struggling to find qualified graduates. One manager who had worked in the field for nearly 20 years, 15 of those in management, said that she had tried to recruit from colleges for years to no avail and wondered if college students were simply not interested in careers in technical communication. She mentioned DITA and SharePoint knowledge as basic requirements and described responsibilities such as creating structured content and UI text, managing content in a source control system, and sharing content across multiple teams. Notably, as one managing director of a content research consultancy pointed out, content-oriented managers are in part to blame for the struggle to find qualified graduates. This is because, he argued, many have set up content ecosystems that are "diabolically hard to work with." As an example, the managing director mentioned a customer who had admitted that their DITA environment had gotten so complex, and the markup so arcane, that it was taking new hires a year to get up to speed, resulting in some of them leaving after just a few months. It is possible that many technical communication students do not apply for content-focused positions because they do not see themselves reflected in the job requirements described in position advertisements, and it is likely that many who do apply and interview are scared away when they learn about an organization's content ecosystem.

In the discussion that followed our session "Building a Community Table" at the 2022 CIDM Best Practices Conference, multiple managers weighed in on the various challenges that they had faced when hiring for entry-level positions. They wanted to hire graduates of technical communication programs, but they had found, in general, that most graduates either did not have a strong technical background or they were uninterested in working in highly technical environments. One manager talked about how she had interviewed two writers who decided not to join her company because they perceived it as "too technical." Another manager talked about how new hires tended to be afraid of the technical aspects of her company, which she had found surprising given that, in her opinion, "technical writers need to have a strong technical background." The technical communication graduates who seemed to be favored most for internships and entry-level positions, at least among the managers who participated in our session discussion, were those who had taken coursework in a STEM discipline.

Graduates of computer science programs were also highly favored, particularly for documentation positions in software development contexts. Deciding to hire someone with a strong technical background versus someone with a strong writing background required, for many managers, carefully weighing benefits and risks. Scott Abel, a well-known content strategist who has hosted dozens of webinars and podcasts with recruiters and managers, spoke to this dilemma. In an August 2021 personal conversation, he said a theme that comes up over and over is that recruiters and managers can't find "technical" writers—meaning people who have diversified skills, not just people who can write but who can also demonstrate technical aptitude, taxonomy knowledge, structure knowledge, XML/HTML/Markdown knowledge, and business savvy, to name a few.

People who say they are technical writers, suggested Abel, tend to be just writers, and people who are not technical writers but more STEM types tend to lack the rhetorical chops.

Hiring preferences for content professionals with technical backgrounds have made it more difficult for managers to find candidates with strong writing skills. In their survey of hiring and training practices in the CIDM community, Stevens and Madison (2021) asked respondents to rank the most important qualifications to find and the hardest qualifications to find. Writing skills ranked as the most important qualification yet the second hardest to find (industry/product knowledge was the first hardest to find). Given the number of students graduating with degrees in technical communication and related fields, from programs that teach writing as a core skill set, this finding certainly surprised us. But read through the lens of our own findings, we know that managers are receiving job applications mainly from candidates with STEM—and particularly computer science—backgrounds; these candidates include both new graduates and people who have come to technical communication as a second career. The technical writing skills that managers are looking for, i.e., structured writing, are rarely represented in the job materials of candidates regardless of degree or prior career. Hence, managers are struggling to find candidates with the kinds of writing skills required for work on technical content teams.

One way some organizations are overcoming hiring challenges is through internship programs and university recruiting efforts. The investment is significant but the payoff, as reported by many managers, is worth it. One survey respondent shared: "Over the past 10 years, my company has put a lot more emphasis on its internship program, university recruiting, and developing early-career talent. My team has greatly benefited from the brilliant early-career writers we have been able to attract." A director of a content solutions consultancy reported that she worked with several companies that had invested in internship programs with good success. They recruit directly from technical communication programs, train interns into their system, and—when they can—offer them full-time positions. Internships are a promising solution to meet the need for a new generation of content professionals and to attract college students to the discipline of content.

Technical Communication Alumni Job Search Experiences

In our interviews with 13 recent alumni of undergraduate programs in technical communication or related fields who had moved to full-time jobs as technical communicators, we asked questions about their career goals and hopes when they had graduated and about their experiences applying for jobs in the field, including for what kinds of jobs they felt qualified and unqualified. We wanted to learn about their struggles and successes, identify any patterns across transcripts that seemed to account for those struggles and successes, and assess the extent to which their experiences aligned with what we had been hearing from managers regarding hiring needs.

Of the 13 alumni, eight applied for entry-level technical writing positions during their senior year (seven of these wanted to pursue a career in technical communication, whereas one did not have a plan). Five of the alumni applied for positions tangentially related to technical communication, such as in federal regulations and e-learning and development. In this section, we feature the eight journeys of the alumni who applied for technical writing positions.

We divided the eight alumni into two groups—those who had training in a technical discipline and/or in structured authoring and those who did not. Five of the alumni did have such training, and all five were working in technical content-focused jobs at the time of the interview (we feature these five alumni—Kat, Nick, Jessica, Lang, and Holly—in Chapter 3). These alumni reported successful job searches; while they had originally sought out any kind of technical writing job, they all landed technical content-focused jobs soon after graduation. One of the alumni working in a technical content-focused job did not have such training and had a less successful job search, having applied for 50–75 positions and receiving one offer. Two of the alumni were working in more traditional technical writing positions, or positions in which they were working with unstructured content using tools such as Microsoft Word and the Adobe Creative Suite. They did not have training in a technical discipline and/or structured authoring, and both reported frustrating job searches.

Alumni with Training in a Technical Discipline and/or in Structured Authoring

These five alumni of programs in technical communication and related fields had completed internships as part of their undergraduate education, four of those internships being in technical writing and one being in communications and marketing. Both Kat's and Jessica's internships turned into full-time positions upon graduation.

All five reported successful job searches in finding entry-level technical writing positions that aligned with their goals. They each attributed their successful job searches to their interdisciplinary backgrounds and knowledge of structured authoring, and particularly the Darwin Information Typing Architecture (DITA) standard. The organizations for which they were working had either adopted digital content transformation workflows or were in the process of doing so.

Table 4.3 presents a summary of the background of these five students when they applied for entry-level technical writing positions.

TABLE 4.3 Backgrounds of Students Who Were Successful in Finding Technical Writing Jobs in the Technical Content Facet of the Field

	Degree/Emphases	*Internships Completed*	*Reported Beneficial Skills & Knowledge*	*Job Title/Company*
Kat	BA in English, emphasis in Professional and Technical Writing AND Literature and Language in Pre-Law Minor in Green Engineering (she was an engineering major for two years)	Yes—technical writing intern with Content Solutions Consultancy first six months after graduation. Hired full time after that	Structured authoring, DITA, engineering	Technical Specialist at Content Solutions Consultancy
Holly	BA in English, emphasis in Professional and Technical Writing Minors in Social Change and Leadership	Yes—marketing and communications intern for three different organizations	Structured authoring, DITA, Markdown	Content Developer at Multinational Technology Company
Nick	BS in Technical Writing and Communication, emphasis in IT Minors in Leadership and Digital Media Studies	Yes—technical writing intern for electronics and smart technology company and marketing intern nonprofit organization	Computer science, programming	Associate Technical Writer at Supercomputer Manufacturer
Lang	BS in Technical and Scientific Communication Minor in French and Francophone Studies	Yes—technical writing intern for machinery company	Computer science, APIs, engineering	Staff Technical Writer at Cybersecurity Software Company
Jessica	BS in Technical Writing and Communication, emphasis in Information Technology and Design Minor in Interdisciplinary Design	Yes—technical writing intern for Fluid Handling Equipment Manufacturer, offered full-time position before graduation	Structured authoring, DITA, information technology, graphic design	Intermediate Technical Writer at Fluid Handling Equipment Manufacturer

Alumni without Training in a Technical Discipline and/or in Structured Authoring

Two alumni of programs in professional and technical communication, Elaine and Don, reported frustrating job searches. Elaine had applied for many jobs but did not get past the second round of interviews. She was able to turn her technical writing internship at the time into a full-time position after not having received any job offers. Don applied for 50–75 positions and received one offer—he had not completed any internship during his undergraduate program. He took the offer. Cole, third alumnus from this group, did not conduct a formal job search, as he felt fortunate to have landed a change analyst internship on the recommendation of another intern who worked at the company. Cole was offered a full-time position after graduation and was grateful to not have had to apply for positions for which he did not feel qualified. At the time of the interview, he had started a master's program in technical communication.

Although these alumni did not describe what they thought had accounted for their job search experiences, we did observe in our analysis of interview transcripts that none of them had taken coursework in structured authoring or a technical discipline. This may have been a factor in them not getting multiple offers or, in the case of Cole, not applying for positions. In their interviews, they each talked about how they wished they had taken courses in areas such as computer science, engineering, and data visualization given the qualifications listed in job ads and given what they had learned in their current positions. Elaine had taken courses in web development and had experience using different programs in the Adobe Creative Suite, and she said those skills had been invaluable in her position at an airline company. Notably, our research on the job market for technical writers suggests that traditional technical writing positions are declining while technical content-focused positions are growing; when graduating seniors review job ads, then, a good number of ads, even though they say "technical writer," are for technical content-focused positions.

Table 4.4 presents a summary of the background of these three students and their job search experience.

Job Search Experience Takeaways

Given that we interviewed only eight alumni of technical communication programs who applied for entry-level technical writing positions during their senior year, we cannot generalize as to why some graduates have more successful job searches than others. But we can highlight here important considerations for industry leaders and hiring managers who are concerned about the aging workforce and want to attract more college graduates to the profession. These considerations also have implications

TABLE 4.4 Backgrounds of Students Who Expressed Frustration with Job Search

	Degree/Emphases	*Internships Completed*	*Reported Frustrations*	*Job Title/Company*
Elaine	BA in Professional and Technical Communication Minor in Public Administration	Yes—marketing intern for a zoo and technical writing intern for a major transportation company; offered full-time position after graduation	Applied for many technical writing positions starting a year before graduation. Did not receive any offers. Stayed on as an intern with the major transportation company for six months after graduation and was offered a full-time position with the company.	Communications and Engagement Specialist at Major Transportation Company
Don	BS in Technical Communication, emphasis in Data Analysis and Presentation	No	Applied for 50–75 jobs in technical writing, graphic design, communication, and advertising. Only heard back from Big Software Developer and was offered the position.	Technical Writer at Big Software Developer
Cole	BA in English, emphasis in Technical Communications	Yes—change analyst intern for a social media corporation; offered full-time position after graduation	Did not apply for technical writing positions because there were not many where he lived. When applying for his internship, he had experience only with Word and PowerPoint and worried he was not qualified.	Technical Document Analyst at Social Media Corporation

for programs wanting to build curricular pathways for students interested in pursuing a career in the discipline of content.

First, the required one to five years of experience listed in most job ads for entry-level technical content positions (most often advertised as "technical writing" positions) discourages many students from applying. Internship experience certainly helps, but many students have not worked in content-related positions for more than a year. Six of the alumni talked about their experiences reading job ads and feeling frustrated by the years of experience required. Jessica and Nick had the following to say:

> I'm a mentor in the advisory board at [my alma mater] for this major, so it's been interesting talking to current students and trying to help them figure out what they feel comfortable applying for because I feel like there's a lot of technical writing jobs in the area, but they always have that barrier of one to five years of experience, and it's like, "Well, how do I get one to five years of experience if I don't have one to five years of experience?"
>
> —*Jessica*

> I guess right off the bat, I was kind of surprised about the amount of experience that some of the entry-level positions were requiring. Which it was like several years of, or like two years of technical writing experience or something. That's something I saw for a lot of associate or tech writing level one positions. So, that was a little difficult to figure out if my experience doing other forms of writing like marketing or web design were applicable.
>
> —*Nick*

In our conversations with industry leaders, some commented on how recent college graduates seemed uninterested in pursuing technical communication as a career given how few applications they had seen from recent graduates. Our own experiences working with undergraduates coupled with what we have learned from alumni tell us that many students don't apply for technical writing entry-level positions because they lack the required one to five years of experience, and they feel unqualified when they see a long list of desired and preferred qualifications that they do not have. Even students who do meet basic qualifications can feel discouraged, as Kat explained when talking about her experience reading job ads:

> I felt unqualified for all of them, but I think that's because I hadn't really had a real job yet. It felt like all of my experience was academic.... I knew I marked off the boxes. Some of them would say structured

> authoring or XML, but I was like, well, I know what that is, and I've touched it, but it's always like one to three years of experience. Well, I did it for one year in college. Does that count?
>
> —*Kat*

She went on to talk about how most college graduates in technical communication do not have all the skills listed in job ads, and few have experience working with structured content—what she knew, based on research she had conducted in college, was now expected for technical content-focused jobs.

If we are to attract a new generation of technical content professionals to the field, those of us working in industry and in academia have work to do. Our colleagues in industry need to reach some consensus to define what an entry-level content-focused position looks like and what experience and qualifications can realistically be expected from college graduates with degrees in technical communication and related fields. Career paths for technical content professionals have yet to be clearly articulated. The corporate ladder would do well to figure out what level 1, 2, 3, and 4 positions look like; this articulation would go a long way in helping trainers and educators develop curricula and students develop academic plans. Leaders in the content industry, particularly those working in the areas of content operations and content strategy, also need to come together to define progression routes for those in entry- and mid-level content-focused positions; a widely shared and agreed upon articulation of advancement paths would go a long way in attracting new and motivating existing talent as well as in providing guidance for trainers and educators interested in developing curricula. Professional organizations such as the CIDM and even smaller groups like Write the Docs might take the lead in articulating career paths and position levels and templating sample job descriptions that hiring managers can adapt.

Hiring managers and program leadership, too, need to create more internship opportunities that afford students windows to gain the experience, skills, and know-how required to succeed in today's highly technical, fast-paced publishing environments. Internships play a critical role in bridging the experience and qualifications gap, and they can provide a steady stream of talent when full-time positions open up. Interns moving into full-time positions will presumably be more equipped to handle the "sink or swim" environment that contributes to entry-level position hiring challenges. For all of the alumni whom we interviewed, with the exception of one, internships were keys to them landing full-time positions after graduation. And finally, as we argue throughout this book, academic programs need to create curricular pathways that not only prepare students for entry-level content-focused positions, but also for advancement into more strategic roles that afford them opportunities to lead and innovate.

In Chapters 5–7, we turn to examining how academic programs might create curricular pathways that prepare a new generation of technical content professionals. While Chapter 5 reviews existing literacy frameworks in technical communication education, Chapter 6 situates within those discussions the skills, competencies, and qualities that industry and academic leaders as well as hiring managers say are necessary for performing knowledge work in the discipline of content. Chapter 7 presents a staged approach to integrating technical content instruction in curricula and shares the advice that both industry and academic leaders had for technical communication educators on how to keep curricula aligned with the needs of the content-focused facet of the field.

References

Abel, S. (2018). Survey reveals top tools, trends, and technologies in use in technical communication teams. *Intercom*, 33–38.

Abel, S. & Bosek, P. (Hosts). (2022, January 20). *Preparing the next generation of technical communication leaders and innovators* [Video podcast episode]. In *Coffee and Content*. The Content Wrangler. Retrieved from https://www.brighttalk.com/webcast/9273/506533.

Andersen, R., & Hackos, J. (2021). Practicing technical communicators' experiences with and perspectives on academic publishing. *Technical Communication, 68*(3), 29–55.

Anderson, P. (1978, April 19–21). *The undergraduate English major with an emphasis in technical writing* [Paper presentation]. The Council for Programs in Technical and Scientific Communication Proceedings, 47–55. https://cptsc.org/wp-content/uploads/2018/01/1978.pdf.

Bosek, P. (Host). (2022, October 13). *The new generation of tech writers* (No. 57) [Audio podcast episode]. In *Content Components*. Heretto. Retrieved from https://heretto.com/project/content-components-podcast/.

Brumberger, E., & Lauer, C. (2015). The evolution of technical communication: An analysis of industry job postings. *Technical Communication*, *62*(4), 224–243.

Carliner, S., & Chen, Y. (2018a). Who technical communicators are: A summary of demographics, backgrounds, and employment. *Intercom*, 8–12.

Carliner, S., & Chen, Y. (2018b). What technical communicators do. *Intercom*, 13–16.

Dweck, C. (2007). *Mindset: The new psychology of success*. New York: Ballantine Books.

Johnson, T. (2018a, October 2). Tech comm trends: Providing value as a generalist in a sea of specialists (Part II) [Blog post]. *I'd Rather be Writing*. Retrieved from https://idratherbewriting.com/2018/10/02/providing-value-as-generalists-in-specialist-contexts-part-2/.

Johnson, T. (2018b, October 2). Tech comm trends: Providing value as a generalist in a sea of specialists (Part I) [Blog post]. *I'd Rather be Writing*. Retrieved from https://idratherbewriting.com/2018/10/02/providing-value-as-generalists-in-specialist-contexts-part-1/.

Lanier, C. (2009). Analysis of the skills called for by technical communication employers in recruitment postings. *Technical Communication*, *56*(1), 51–61.

Masycheff, A. (2018). Keeping technical writing jobs relevant with content automation technologies. LinkedIn. Retrieved June 10, 2022 from https://www.linkedin.com/pulse/keeping-technical-writing-jobs-relevant-content-alex-masycheff/.

Melonçon, L. (2009). Master's programs in technical communication: A current overview. *Technical Communication*, *56*, 137–148.

Melonçon, L. (2012). Current overview of academic certificates in technical and professional communication in the United States. *Technical Communication*, *59*, 207–222.

Melonçon, L. (2014). Curricular challenges of emphasis degrees in technical and professional communication. In T. Bridgeford, K. Kitalong, & B. Williamson (Eds.), *Sharing our intellectual traces: Narrative reflections from administrators of professional, technical, and scientific communication programs* (pp. 179–200). Amityville, NY: Baywood.

Melonçon, L. (2021). TechComm Programmatic Central (unpublished raw data). Retrieved from https://tek-ritr.com/techcomm-programmatic-central/.

Melonçon, L., & Henschel, S. M. (2013). Current state of U.S. undergraduate degree programs in technical and professional communication. *Technical Communication*, *60*(1), 45–64.

Schengili-Roberts, K. (2022a, August 29). Now is the best time to be a technical writer (despite a possible oncoming recession [Blog post]. *<dita>Writer*. Retrieved from https://www.ditawriter.com/now-is-the-best-time-to-be-a-technical-writer-despite-a-possible-oncoming-recession/.

Schengili-Roberts, K. (2022b, October 26). Now is no longer the best time to be a technical writer looking for a job [Blog post]. *<dita>Writer*. Retrieved from https://www.ditawriter.com/now-is-no-longer-the-best-time-to-be-a-technical-writer/.

Stanton, R. (2017). Do technical/professional writing (TPW) programs offer what students need for their start in the workplace? A comparison of requirements in program curricula and job ads in industry. *Technical Communication*, *64*(3), 223–236.

Stevens, D., & Madison, K. (2021). Hiring and training: What's happening in the CIDM community. *Best Practices Newsletter*, 10–17.

Write the Docs. (2023). Documentation salary survey 2023 results. Retrieved from https://www.writethedocs.org/surveys/salary-survey/2023/.

SECTION III

Competencies, Curricular Pathways, and Programmatic Innovations

5

TECHNICAL COMMUNICATION LITERACY FRAMEWORKS AND APPROACHES TO IDENTIFYING COMPETENCIES AND SKILLS

Strengthening Connections between Technical Communication Education and Practice

Industry Leaders Want to Partner with Educators

Our interest in examining the implications of digital content transformations for education in technical communication stemmed in part from various conversations that we had been having with leaders in industry at a number of professional conferences.

Since 2010, Rebekka has been a regular participant in the annual Center for Information-Development Management (CIDM) Best Practices Conference, an event that brings together leaders and managers of content organizations and teams. Several presentations each year focus on the changing workforce and staffing needs, with presenters and attendees often discussing challenges around hiring and asking questions about how college graduates are being prepared for new roles on technical content teams. At the 2013 conference, for example, the panel "Building a Staffing Model for the Future" posited a staffing model "for the new normal, which is a fast paced, fluid, flexible, constantly changing environment." In this new normal, "writers need to be knowledge engineers" and "SMEs [Subject Matter Experts]" and engage in "extreme levels of collaboration." Panelists spent a good deal of time describing staffing problems in their respective organizations and gaps in the skills and expectations of job applicants. The 2014 conference similarly featured a keynote panel focused on the changing workforce. In addition to describing the skills

DOI: 10.4324/9781003451730-9

most needed, panelists pondered as to why there seemed to be so few new people in the field.

As the only academic participant most years, Rebekka would often field questions from attendees who were curious about curricular offerings in technical communication programs and ways in which programs were preparing students for entry-level positions in the content discipline (and more specifically, on teams that had adopted structured authoring workflows). Industry folks genuinely wanted to figure out how they could best attract and onboard through training and mentoring new content professionals. And some wanted to know how they could collaborate with programs, from contributing course design ideas and materials to giving guest lectures to participating on advisory boards.

Carlos, too, as a member of the DITA Technical Committee in the mid-2010s and co-chair of the Lightweight DITA subcommittee, had been fielding similar questions from leaders who had been instrumental in defining the widely adopted content standard. Some of these conversations led to curricular collaborations, such as one between Carlos and Michael Priestley—one of the creators of the DITA standard for technical content. Carlos and Michael (Evia & Priestley, 2016), who were working on the Lightweight DITA standard for technical documentation, designed an assignment in a technical writing class that asked students to author and publish instructions with HDITA, an HTML5-based version of DITA. They analyzed students' experiences in the different stages of the HDITA authoring process and evaluated the extent to which these novice technical writers could author effective topic-based technical content and do so with relative ease. This collaboration was a direct attempt to bridge industry practice and curricula with the aim of preparing students for entry-level positions in technical content authoring environments.

In the mid-2010s, our conversations made it clear that a growing gap was forming between what industry was doing with technical content and what our undergraduate programs were teaching. This widening disconnect concerned both us, as educators, and industry leaders who wondered where the next generation of skilled content professionals would come from.

When Rebekka visited the Virginia Tech campus in 2016 to give a talk based on her article "Rhetorical Work in the Age of Content Management" (Andersen, 2014), which was the outcome of several years of learning from industry leaders at industry conferences, we decided that it was time to pursue a more systematic approach to understanding the changing roles and activities of technical content professionals and the implications of these changes for education and training in the field. As we mention in the "Introduction," this systematic approach became the multi-year research project that culminated in this book.

In the early stages of the project, we were invited to collaborate with industry leaders on panels focused on the "State of Education in Technical Communication" at the 2017 CIDM Best Practices Conference and a 2018 Data Conversion Laboratory webinar. At the time, the three industry leaders on the panels were each teaching part-time in technical communication undergraduate or university extension programs, and they were each working full-time as either managers or consultants. While our role on the panels was to describe the landscape of education programs and curricula and ways academia and industry could work together, the other panelists' role was to describe the environments into which college graduates were being hired and the friction points between those environments and the preparation strengths and deficiencies of graduates. One of those panelists also described how they were working to bridge the skills gaps through a robust onboarding program, and another described how he had begun to develop a clearing house for shared instructional content to make recent graduates more marketable and industry onboarding less expensive—he used that opportunity to invite attendees to contribute to the initiative.

Events such as these point to opportunities for program administrators and faculty to partner with industry leaders in the content discipline to build curricular pathways that prepare students for work as content professionals. These kinds of collaborations are essential for identifying ways in which practice and education at the intersection of technical communication and the discipline of content might come into better alignment.

Our co-edited 2018 special issue of the journal *Communication Design Quarterly* (Andersen & Evia, 2018), focused on "Perspectives on Preparing Technical Communication Professionals for Today and the Future," included several articles that argued for and demonstrated the value of innovating curricula in collaboration with industry peers (see Carnegie & Crane, 2018; Duin & Tham, 2018; Shalamova, Rice-Bailey, & Wikoff, 2018). Nadya Shalamova, Tammy Rice-Bailey, and Katie Wikoff (2018), for example, argued that "partnering and maintaining ongoing relationships with local employers of technical communicators" is key to the field's ability to evolve and innovate to remain relevant (p. 19). Teena Carnegie and Kate Crane (2018), too, stressed the importance of developing "curriculums that seek to reflect the tasks and activities, competencies and skills required by the workplaces that will employ their students" (p. 11). Bringing together industry and academic perspectives on educating and training technical communication professionals for today and the future, the special issue underscored the need for technical communication education to stay connected to practice. Fast forward to the 2020s, with digital content transformations redefining technical content work, this connection is more important than ever, at least for programs wanting to develop curricular pathways that prepare students for entry-level (and beyond) content-focused jobs.

Educators Want to Align Curricula with Practice

Arguments for keeping our educational programs connected to professional practice are certainly not new to the field (Figure 5.1 illustrates the importance of this connection). Technical communication's academic identity is largely rooted in the connection between practice and education. Carolyn Rude (2015) suggested that technical communication developed its sense of identity in the 1980s and 1990s in large part by developing programs and curricula aimed at preparing students for careers in industry. Stan Dicks (2002) and Miles Kimball (2015), too, pointed back to those roots:

> Academic programs in technical communication came into existence because of the needs of business and government for competent communicators with special knowledge and skill sets required to produce technical documents of high quality. Those programs, in a real sense, depend for their existence on continuing to meet the needs of the "work world."
>
> *(Dicks, 2002, p. 13)*

> Suffice it to say that the bridge between school learning and professional practice has long been a matter of concern for researchers and teachers, who have expended considerable time and effort to find out what companies need and want in a technical communicator, and to bring curricula into alignment with those requirements and desires.
>
> *(Kimball, 2015, p. 136)*

As the field has evolved and diversified, researchers and teachers have continued to find ways to align curricula with the work and needed skill sets of industry-situated technical communicators (see, for example, Blakeslee, 2001; Carnegie & Crane, 2018; Duin & Tham, 2018; Kimball, 2015; Newmark & Bartolotta, 2021; Shalamova, Rice-Bailey, & Wikoff, 2018). Julianne Newmark and Joseph Bartolotta (2021), for example, reformed a foundation-level technical writing course's learning objectives based on the nine areas of competency that comprised the Society for Technical Communication's (STC) Foundation Certification Exam. They wanted to know if there were "ways to reformulate a curriculum, whole cloth, so that the founding premises of that curriculum have an eye towards students' eventual workplace realities" (p. 148). Their research took them to nine competency areas, which they embedded as an "enculturative framework" throughout the undergraduate technical and professional communication (TPC) curriculum. Using the competencies to define learning objectives throughout the curriculum offered students a vocabulary for talking about their knowledge and specialized skills to employers and, although not the same as a professional certification, lent credibility to their educational preparation.

As activities, roles, and content ecosystems have become more complex in the content-focused facet of our field's collective identity, however, aligning curricula focused on technical content has become more challenging, in part because keeping abreast of trends and making sense of what they mean for educating future content professionals is an increasingly difficult task.

In this chapter, we summarize scholarship in the field that has examined both practice and educational curricula for the purpose of identifying the

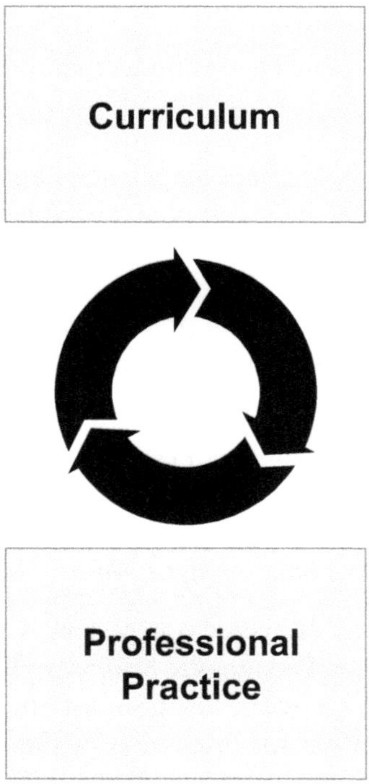

FIGURE 5.1 Researchers and Teachers Have Argued for the Need to Align Curriculum with Professional Practice. However, maintaining this alignment is an ongoing process that requires resources from both sectors of the field involved.

skills, competencies, and qualities that undergraduate programs in technical communication should strive to cultivate in students so as to prepare them for workplace realities. In Chapter 6, we situate within these already codified frameworks the skills, competencies, and qualities that our research participants identified as necessary for performing knowledge work in the discipline of content.

Existing Literacy Frameworks and Approaches to Identifying Competencies and Skills

In this section, we examine how technical communication scholars have identified and described needed competencies and skills, particularly those that connect with practice, for students completing undergraduate degrees in TPC.

Scholars interested in identifying competencies and skills needed for work in different facets of the field more generally have taken two approaches:

1 Examining workplace practices and prior literacy frameworks to update or develop new frameworks.
2 Analyzing job advertisements or conducting interviews or surveys with practitioners.

Scholars interested in identifying competencies and skills specific to work within organizations that have adopted structured authoring workflows have taken a third approach:

3 Synthesizing the content-focused practitioner discourse focused on necessary roles and activities and highlighting the skill sets necessary for carrying out those roles and activities.

  - In the following sections, we describe each of these approaches to identifying competencies and skills.

Layered Literacies and Symbolic-Analytic Work

Since the publication of "Layered Literacies: A Theoretical Frame for Technical Communication Pedagogy," Kelli Cargile Cook's (2002) layered literacies pedagogical framework has been a dominant model in the field of technical communication for discussion of the competencies and skills that students need to succeed in the workplace. The framework identified the following six key literacies for technical communicators:

1 Basic
2 Rhetorical

3 Social
4 Technological
5 Ethical
6 Critical.

As literacies are not isolated, Cargile Cook urged instructors to teach them as integrated, or layered within the curriculum, and to think of them as "layers of learning that move and flow over one another depending on the topic of the moment, hour, unit, or course" (p. 24). The framework was, at the time, in part a response to concerns about needing to prepare students for work in increasingly complex information ecologies.

Twelve years later, inspired by Cargile Cook's layered literacies framework and curious as to how discussions of competencies and skills had evolved, Sally Henschel and Lisa Melonçon (2014) reviewed the body of scholarship that identified conceptual and practical skills needed to be a successful technical communicator. In "Of Horseman and Layered Literacies: Assessment Instruments for Aligning Technical and Professional Communication Undergraduate Curricula with Professional Expectations," the authors addressed the question, "What conceptual and practical skills are deemed important by academics and practitioners?" with the goal of offering educators guidance on applying these skills to course development and program assessment. The authors argued, "Because TPC is an applied field, students who possess both conceptual and practical skills are better prepared to succeed in the workplace" (p. 5).

Scholars over the years have tended to talk about conceptual skills as "knowing that" and practical skills as "knowing how" (see, e.g., Carter, 2007; Henning & Bemer, 2016, Hovde, 2022). Whereas conceptual skills "encompass critical thinking and problem solving, the higher-order knowledge and literacies a technical communicator needs to be successful and remain flexible in the ever-changing workplace," practical skills encompass "specific, identifiable skills" such as "audience analysis, writing, editing, information and document design, and technology/tool knowledge" (Henschel & Melonçon, 2014, p. 5).

Henschel and Melonçon, as well as multiple other scholars, have linked conceptual skills to what former US Secretary of Labor Robert Reich called "symbolic-analytic work," a process that Johndan Johnson-Eilola (1996) described as using skills in "abstraction, experimentation, collaboration, and system thinking to work with information across a variety of disciplines and markets" (p. 248). Reich (1992), in his book, *The Work of Nations*, introduced these higher level skills as those necessary for work in a global, information-driven economy. Since then, several scholars in the field have argued for rethinking technical communication in terms of these skills to move the field away from its focus on production to a broader focus on the knowledge that technical communicators bring to organizations (Durack, 2003; Hart-Davidson, 2001; Johnson, 1998; Johnson-Eilola, 1996; Slattery, 2005; Wilson, 2001). Symbolic-analytic workers, Johnson-Eilola argued, are "highly skilled in information manipulation and abstraction, critical and much sought-after skills in an age where information overtakes industry in terms of social and economic value" (p. 253).

As far back as 1996, Johnson-Eilola urged the field to elevate technical communicators as symbolic-analytic workers who might engage in activities such as usability research; working on distributed teams; and finding and articulating patterns, structures, and relationships in large amounts of information and across specific problems, projects, and task domains (pp. 258–261). Five years later, Greg Wilson (2001), too, drew on Reich's articulation to identify workplace skills necessary for elevating technical communicators as respected rhetorical agents in the 21st-century workplace. By teaching these higher order skills, Wilson suggested, instructors can "imbue our students with the agency to interpret and negotiate both complex information and complex work contexts" (p. 97).

In an early effort to begin mapping how these skills relate to those needed for work in emerging technical content ecologies, Rebekka (Andersen, 2014), drawing on Reich (1992) and Johnson-Eilola (1996), offered the following summaries of skills in abstraction, systems thinking, experimentation, and collaboration:

- *Abstraction*—Skills in abstraction involve being able to solve problems in an innovative, proactive manner. Reich (1992) described abstraction as "searching for new ways to represent reality which will be more compelling or revealing than the old" (p. 229). Symbolic analysts applying such skills, Johnson-Eilola (1996) suggested, "discern patterns, relationships, and hierarchies in large masses of information" (p. 260).
- *Systems thinking*—Systems thinking involves the ability to understand how different parts of the whole are related. When presented with a problem or an issue, Reich (1992) described, systems thinkers can "discern larger causes, consequences, and relationships" (p. 30); a symbolic analyst can redefine problems in terms of a "broad system of forces, variables, and outcomes" that might not be immediately apparent (p. 31).
- *Experimentation*—Skills in experimentation entail the ability to try out new techniques and explore a range of possibilities and outcomes (Reich, 1992, p. 31); experimentation "involves forming and testing hypotheses about information and communication" (Johnson-Eilola, 1996, p. 258).
- *Collaboration*—Skills in collaboration involve being able to share knowledge, ideas, problems, and solutions in an effort to reach consensus

> on the best plan for moving forward (Reich, 1992, p. 32). Symbolic analysts can articulate clearly their ideas to others. They can "accept criticism from peers, solicit help, and give credit to others," and they can "negotiate—to explain their own needs, to discern what others need and view things from others' perspectives, and to discover mutually beneficial solutions" (p. 32).

We want to call particular attention to ways in which these higher order skills have been linked, from as early as 1992, to work in complex information systems and environments. Systems, technologies, processes, and contexts, while complex 30 years ago, are exponentially more complex today in organizations that have adopted digital content transformation workflows. And thus technical communication educators need a way to make sense of what these higher order conceptual skills look like in modern content ecosystems and what "how-to" practical skills are needed to perform knowledge work in conjunction with those conceptual skills.

Henschel and Melonçon (2014), valuing the utility of Reich's higher order skills for symbolic-analytic work and Cargile Cook's (2002) layered literacies, saw commonalities between these articulations of "conceptual skills," despite them being developed for different purposes: "Reich's from an industry standpoint and Cargile Cook's from an educational one" (p. 7). As such, they decided to combine some of Cargile Cook's literacies (basic and rhetorical, ethical, and critical) and then compare the combined literacies with the four pillars of symbolic-analytic work. In doing so, they were able to identify five conceptual skills that they argued represented "the higher-order knowledge and literacies a technical communicator needs to be successful and remain flexible in the ever-changing workplace" (p. 5). Henschel and Melonçon represented these five conceptual skills in an explanatory matrix meant to help instructors align curricula with professional expectations.

Figure 5.2 presents the explanatory matrix created by Henschel and Melonçon, reproduced and used with permission here. Conceptual skills deemed important are represented by four quadrants anchored by *critical system thinking*: these skills include *rhetorical proficiency*, *abstraction*, *experimentation*, and *social proficiency*. The authors note,

> By tethering the quadrants together with critical system thinking, we acknowledge programmatic goals to help students become critical system thinkers who can deploy any one or a combination of the conceptual skills as needed to solve problems in the workplace.
>
> *(p. 9)*

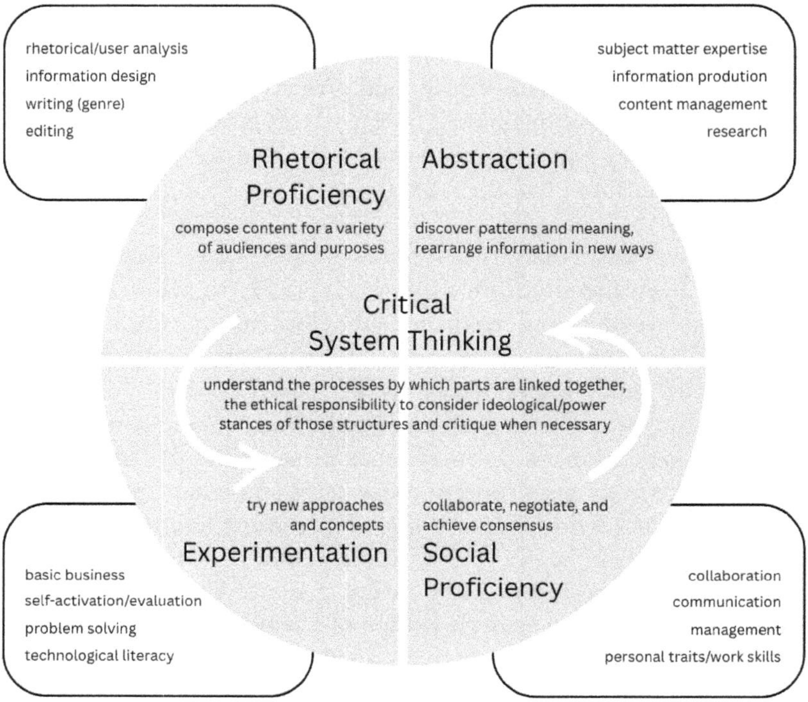

FIGURE 5.2 Henschel and Melonçon's Explanatory Matrix of Conceptual and Practical Skills for Technical Communicators. Reproduced and used with permission. Image credit: "Of Horsemen and Layered Literacies: Assessment Instruments for Aligning Technical and Professional Communication Undergraduate Curricula with Professional Expectations," by S. Henschel & L. Melonçon, 2014, *Programmatic Perspectives*, 6(1), 13.

As Reich's descriptions of symbolic-analytic work were not focused on communication tasks, he did not account for rhetorical agility in his descriptions of higher level skills. Henschel and Melonçon thus retained Cargile Cook's rhetorical literacy in their matrix (called *rhetorical proficiency* in their model), as it accounts for the higher level work of shaping communications for particular audiences and purposes. The authors identified *abstraction* as a new conceptual skill not represented in the layered literacies framework, and they combined Cargile Cook's social literacy and Reich's collaboration into the conceptual skill of *social proficiency*. Viewing technological literacy as one of many facets of experimentation, the authors decided to add *experimentation* as a conceptual skill in the matrix and list technological literacy as a practical skill connected to it.

Critical system thinking, as the conceptual skill that anchors all others, encompassed both Cargile Cook's critical literacy and Reich's systems thinking skills—Henschel and Melonçon saw these as "extensions of one another: to be able to recognize and consider ideological stances (Cargile Cook), one needed to be able to see the whole and how it was linked together (Reich)" (p. 8).

The boxes connected to each quadrant list practical "how-to" skills in relation to conceptual skills. Henschel and Melonçon examined eight studies that identified practical skills deemed important (some conducted by academics, others by practitioners) and noted the most often recommended practical skills. They then located them on the matrix, explaining that "The quadrants and corresponding practical skills illustrate the relationship between each part to the whole—a way of seeing what the discipline deems important…" (p. 12).

We find Henschel and Melonçon's matrix helpful in thinking through skills and competencies that map onto many work contexts, and we find it particularly useful for thinking through competencies needed to understand, contribute to, and innovate within complex information systems, including those connected to digital content transformations. Both the conceptual and practical skills on the matrix can be adapted and expanded upon to reflect those needed for different technical communication positions in different work contexts and, for our purposes, those needed for content professional roles in organizations that have adopted structured authoring workflows. We particularly appreciate how the matrix as a literacy framework attends to the need to cultivate graduates—critical pragmatic practitioners—who can readily move from entry-level positions into more strategic positions.

One alternative view of conceptual and practical skills that we wanted to note is one where technological literacy might be located on the explanatory matrix. Whereas Cargile Cook identified technological literacy as a conceptual skill, Henschel and Melonçon did not elevate it as its own conceptual skill category in the matrix; rather, they identified it as a practical skill within the category of experimentation. Others, including Cargile Cook, however, have described *technological literacy* more along the lines of a conceptual skill (see, e.g., Brumberger, Lauer, & Northcut, 2013; Duin & Tham, 2018; Hovde, 2022; Hovde & Renguette, 2017; Selber, 2004). Given the growing importance of this literacy for work in changing content ecosystems, we feel it important to note others' rationales for doing so.

Marjorie Rush Hovde (2022) argued for a view of technological literacy as both a practical skill and conceptual skill, or what she called "procedural knowledge." Procedural knowledge, she proposed, requires both "how-to" skills and "conceptual knowledge of a system" (p. 138). How-to skills make following routine processes or working within simple systems easy. But they become limited when contextual factors disrupt routine processes and require considerations of alternatives. They become even more limited when processes are connected to complex systems. Conceptual knowledge is thus additionally necessary, as it "provides an important foundation when tackling non-routine and/or complex processes" (p. 140). Applying critical thinking and problem-solving to non-routine situations, technical communicators draw on conceptual knowledge to understand what can be done and why and then decide how best to do it given constraints and resources. Technological literacy in this view is not about knowing particular tools but about knowing how to navigate messy situations and move between and among technologies and tools to problem solve.

In an earlier article, Marjorie Rush Hovde and Corinne Renguette (2017) made a strong case for elevating technological literacy to a conceptual skill necessary for performing well in the workplace. In support of their case, they point to the definition of technological literacy offered by the International Technology Educational Association (2006)—"'the ability to use, manage, evaluate, and understand technology' (p. 4) to perform a wide variety of functions"—and to the work of Littlejohn et al. (2012), who argue that "technologically literate users transfer skills to multiple contexts and applications" (p. 396). These definitions situate technological literacy within the realm of critical thinking and problem-solving, what Henschel and Melonçon described as the "higher-order knowledge and literacies a technical communicator needs to be successful and remain flexible in the ever-changing workplace" (p. 5).

Hovde and Renguette define four levels of technological literacy that technical communicators need to succeed in the workplace. These levels include the following:

- *Functional literacy* (how-to, or effective and efficient use of technology)
- *Conceptual literacy* (understanding of a technology and the larger contexts in which it facilitates work)
- *Evaluative* (ability to problem solve through appropriate tool selection)
- *Critical* (understanding technology impacts).

Technological literacy defined in terms of levels includes both conceptual skills and practical skills, in much the same way as Hovde's (2022) description of procedural knowledge does. One criticism of frameworks such as that proposed by Hovde and Renguette (2017), however, is that they tend not to address code literacy. Seeing this as an opportunity, Ann Duin and Jason Tham (2018) advocate for a greater focus on code literacy within the four-part framework developed by Hovde and Renguette.

Technical communicators, they argue, need to understand "specific authoring languages for use in organizing information to communicate or execute actions"; such languages include standards like HTML, XML, and DITA. The authors break code literacy into three facets—code-as-language, code-as-tool, and code-as-structure—that can be mapped onto the four levels of technological literacy.

Drawing on Henschel and Melonçon's explanatory matrix and discussions of technological literacy as a conceptual skill that also includes code literacy, we move in Chapter 6 to expand the matrix to map with the discipline of content. We do this after presenting what experts, managers, and alumni had to say about what they view as the most important skills and competencies for students completing degrees in the field to learn. We hope educators and administrators will find the expanded matrix, which we introduce as a competencies framework for content professionals, useful for the purpose of planning program and curriculum innovations.

Importantly, as we conducted focus groups and interviews for our research project, we kept these questions in mind:

- What do the five conceptual skills in Henschel and Melonçon's matrix look like in practice? How do new graduates apply these skills in today's content environments?
- What practical skills are missing from the matrix? What knowledge areas do these skills encompass?
- What does technological literacy look like for content professionals working in the disciplinary areas of content and particularly for entry-level content professionals working in the area of content design as technical content developers?

Analyses of Job Advertisements and Interview and Survey Reports

Whereas layered literacy frameworks such as those offered by Cargile Cook (2002) and Henschel and Melonçon (2014) have offered educators a bird's-eye view approach to understanding competencies needed to perform well in the technical communication workplace, studies that report results from analyses of job advertisements and from interviews and surveys of practitioner leaders have offered educators an on-the-ground approach to understanding both required and desired competencies (examples, among others, include Brumberger & Laeur, 2015; Kimball, 2015; Lanier, 2009; Shalamova, Rice-Bailey, & Wikoff, 2018; Stanton, 2017). So too have reports of competencies published by leaders in the field formerly associated with the defunct STC; these leaders have drawn on their wealth of practitioner experience, conversations with STC members, and workplace research to identify core competencies for technical communicators. These kinds of studies and reports amplify voices from industry, particularly those who regularly hire entry-level technical communicators, and they play an important role in helping educators keep undergraduate programs connected to professional practice.

With the exception of a handful of studies (notably, Lanier, 2009 and Stanton, 2017, which examined only "technical writer" job ads), analyses of job advertisements and interview and survey reports to date have focused on the field of technical communication at large, offering assessments of a broad sweep of technical communication positions in various industries. As we know from previous research (e.g., Carliner & Chen, 2018; Melonçon & Schreiber, 2022; Shalamova, Rice-Bailey, & Wikoff, 2018), technical communicators assume various roles and titles and work in a variety of communication contexts, creating and designing a wide range of content types and experiences using a wide range of tools. Results of these kinds of studies, then, do not necessarily reflect the evolving roles, activities, and needed skill sets of technical communicators working in technical content-focused positions. Nevertheless, they do help us to understand how curricula, for many technical communication and related undergraduate programs, have evolved to align with industry needs with the goal of preparing students for different career paths within the field. Scholars who have developed layered literacies frameworks have also drawn on these kinds of studies to identify competencies that we see reflected in curricula.

Aggregate Reports of Competencies

Not surprisingly, leaders in the field formerly associated with the STC have contributed a great deal to our collective understanding of competencies needed for the technical communication workplace. In an oft-cited editorial, George Hayhoe (2000), drawing on his extensive experience as editor of *Technical Communication* and STC fellow, asked, "What is the essential knowledge that all technical communicators must possess?" He then offered a breakdown of skills all technical communicators need regardless of specific job. These foundational skills included the following:

- *Knowledge of communication concepts and techniques*—"writing, editing, visual communication, multimedia, document design, audience and task analysis, usability testing of products and documents, and interpersonal communication" (p. 151).
- *Knowledge of technical subject domains*—the need for subject-matter expertise in "the sciences, medicine, engineering, or another technical field" (p. 152).
- *Knowledge of software tools.*

One could categorize these skills as rhetorical proficiency, subject-matter expertise or domain knowledge, and technical knowledge.

Ken Rainey, also a former STC fellow, dedicated much of his academic career to identifying technical communication competencies, drawing on the wealth of knowledge and experience of STC members, and validating his findings through extensive workplace research. In 2005, he and co-authors published the study "Do Curricula Correspond to Managerial Expectations" (Rainey, Turner & Dayton), in which they reported the results of both a survey and interviews of technical communication managers that aimed to identify the most important competencies for technical communicators. Like Hayhoe, they identified rhetorical proficiency and technical knowledge as important competencies but added collaboration and initiative to the list:

- Skills in collaborating with both subject-matter experts and coworkers
- Ability to write clearly for specific audiences directed by clearly defined purposes
- Ability to assess and learn new technologies
- Ability to take the initiative (be a self-starter) and to evaluate one's own work and the work of others.

(p. 323)

Other competencies, which they called "secondary," included "skills in using technologies to accomplish documentation work in various media and the ability to write, edit, and test various technical communication documents" (p. 323). A third level of competencies, which they called "tertiary," included "skills in usability testing, single-sourcing and content management, instructional design, budgeting, oral presentations, research, multimedia, and awareness of cultural differences" (p. 323). These important, secondary, and tertiary competencies, in addition to those identified by Hayhoe (2000), combine both conceptual and practical skills and speak to the ways in which technical communicator activities were changing in the early 2000s, particularly with respect to increased access to new forms of media and early adopters of structured authoring workflows—primarily for the purpose of single-sourcing content.

More recently, the STC worked to codify the work of leaders such as Hayhoe and Rainey and those who had contributed to the Technical Communication Body of Knowledge by defining nine core competencies that served as the foundation for its certification program. Launched in 2011, the program aimed to "create preference in the marketplace for certified technical communicators and to validate efforts in learning core skills, best practices, and specific industry standards" (Pohland & Lyons, 2016, p. 8).

Those aiming to become a Certified Professional Technical Communicator were assessed on the competencies described below[1]:

1 *Project planning*—"focuses on the work involved in planning and managing technical communication work teams and documents through a lifecycle process" (Baehr, 2016, p. 10)
2 *Project analysis*—"involves the work of identifying readers and document contexts, including the development of reader profiles" (Baehr, 2016, p. 10)
3 *Content development*—"covers the tasks, methods, approaches, and practices of: Choosing an appropriate technical genre for a rhetorical situation"; "following an appropriate pattern, as expected by the reader, for each technical genre"; and "conducting research for content development" (Houser, 2017a, p. 9)
4 *Organizational design*—"focuses on guidelines and techniques for organizing and drafting technical documents" (Baehr, 2016, p. 10), including "organizational patterns and rhetorical moves for introductions and conclusions" (Baehr, 2017, p. 11)
5 *Written communication*—"knowledge and understanding of concepts and techniques for developing clear, effective written communication, that also meets requirements for global audiences and translation" (Houser, 2017, p. 13)
6 *Visual communication*—focuses on "methods for designing information to guide your readers" (Gillenwater, 2017, p. 15) and "on general editing principles and practices including using graphics data displays and other kinds of information graphics, such as bar charts, line graphics, tables, pie charts, flow charts, etc." (Baehr, 2016, p. 11)
7 *Reviewing and editing*—emphasizes review and editing processes and guidelines and focuses on the role of usability testing in the review and editing process (Brown-Hoekstra, 2017, p. 18)
8 *Content management* (CM)—"focuses on teamwork and using websites to develop and manage information products"; this competency "ensures projects are delivered to specification, on time, and that supporting tools such as websites and social media contain appropriate content" (Agnew, 2017, p. 21)
9 *Production and delivery*—focuses on producing and delivering information products that meet business goals, that "match the intended outcomes that were identified at the beginning" of a project (Agnew, 2017, p. 24).

Although digital content transformations were underway with early adopters in the 2010s, the core competencies comprising the STC certification program did not fully reflect the expanded skill sets needed for technical content-focused positions. For example, the *content development* and *written communication* competencies focused heavily on traditional technical genres such as reports and on unstructured documentation that was not the output of structured authoring workflows; similarly, *CM* focused more on managing website content than on managing modular, structured content that could be published or served to a variety of outputs and formats. The competencies did, however, reflect activities and processes that cut across job titles, work contexts, and industries, including those connected to the discipline of content.

Building on this earlier work, Yvonne Cleary (2021), in *The Profession and Practice of Technical Communication*, offers the most recent taxonomy of key competencies needed for the technical communication workplace. Cleary reviewed an extensive body of work that identified competencies and skills essential for performing well in the technical communication profession. The body of work included international research projects focused on building competence frameworks, professional association resources, prominent industry blogs, and peer-reviewed quantitative and qualitative studies. Based on her review, she generated a comprehensive list of 14 often cited competencies and skills (see her Table 4.1, p. 88). The list does not distinguish between competencies and skills, though Cleary herself defined competency as "a capacity, such as problem solving, time management, or analytic ability, that you need to be able to do a job well" and skill as "a more specific ability, such as the ability to use a tool, to code, or to write in a particular style" (p. 86). Her definition of competency aligns with others' definitions of conceptual skills, and her definition of skills aligns with others' definitions of practical skills.

For ease of reference and for the purpose of understanding these 14 competencies and skills in the context of a layered literacies framework, in Table 5.1, we grouped the competencies and skills in Cleary's list into the five conceptual skill areas proposed by Henschel and Melonçon (note that we do not distinguish between conceptual and practical skills in Cleary's list).

Some competencies and skills easily fall into two or more conceptual skills areas. *Information development*, for example, is described as "The ability to produce content that is suitable for its purpose and usability for the intended audience" (Cleary, 2021, p. 88); in structured authoring workflows, information development requires skills in not only *rhetorical proficiency* but also *abstraction* and *critical system thinking*. *Management*, too, might just as easily fall under *critical system thinking* as it does *social proficiency*, given that competency in management speaks to the "ability to plan, organize, and steer various elements of your work, including content, people, projects, quality, and tasks" (p. 88).

Cleary also offers a list of transferable skills gleaned from her review of the body of work examining essential competencies and skills in the profession. These transferable skills align with what others have called soft skills or personal characteristics (e.g., Brumberger & Lauer, 2015). The

TABLE 5.1 Competencies and Skills Identified by Cleary (2021) Grouped into the Five Conceptual Skill Areas Proposed by Henschel and Melonçon (2014)

Henschel and Melonçon's 5 Conceptual Skill Areas	*Cleary's 14 Competencies and Skills*
Rhetorical proficiency	• Writing and rhetoric • Revision and review • Audience analysis • Information development • Visual design • Information design • Legal and standards knowledge
Social proficiency	• Communication (oral, interpersonal, intercultural, collaboration) • Management (project management, time management, quality management, content management, people management)
Experimentation	• Technological literacy • Domain knowledge
Abstraction	• Content strategy and information architecture • Research
Critical system thinking	• Usability and user experience

skills include *critical thinking*, *multitasking/task orientation*, *organization*, *detail orientation*, *problem-solving*, and *creativity* (pp. 89–90). Critical thinking and problem-solving are higher order skills that cut across the five areas of conceptual skills in Henschel and Melonçon's explanatory matrix—arguably landing in critical system thinking, which anchors the four other areas; the other transferable skills are more attuned with what we would call essential qualities for performing well.

Single-Study Reports of Competencies

Several scholars since 2015 have conducted studies in the form of analyses of job advertisements or interviews and surveys of US practitioners to identify required and desired competencies for technical communicators. These studies have tended not to distinguish between conceptual and practical skills, but some have identified required and desired "hard" skills versus "soft" or "transferrable" skills. Hard skills, in many cases, can be linked to practical skills while soft skills can be linked to conceptual skills. Brumberger and Lauer (2015), in their study of nearly 1,000 job advertisements, grouped skills into the two categories of professional competencies and personal characteristics; while their list of professional competencies fall under both conceptual and practical skills, their list of personal characteristics are more attuned to the qualities needed to apply conceptual skills well.

In Table 5.2, we present studies published since 2015 that have examined job advertisements or the perspectives of practitioners, particularly those in leadership roles, via interviews and surveys. These studies represent different approaches to identifying needed skills and competencies for technical communicators, with the aim of bringing academic curricula into better alignment with professional practices (the skills and competencies reported in Table 5.2 are listed in order of most mentioned). We did not include studies published prior to 2015, as we wanted to capture lists of competencies relevant to evolving roles and activities of technical communicators. We do note that Clint Lanier's (2009) analysis of technical writing job advertisements served as a methodological framework for Brumberger and Lauer's (2015) as well as Ronda Stanton's (2017) analysis. Other notable studies reporting results of interviews and surveys include Aimee Whiteside (2003) and Barbara Giammona (2004).

Skills most mentioned among the six studies presented in Table 5.2 included the following (we group them here into associated conceptual skill areas):

- *Rhetorical proficiency*—e.g., written communication, technical writing, editing, audience analysis, information design, document design
- *Social proficiency*—e.g., collaboration, working in teams, motivation to help others
- *Technological literacy*—e.g., general software knowledge, technical knowledge, technical acumen
- *Communication*—e.g., oral communication, interpersonal communication, communication strategy
- *Management*—e.g., project management, time management, multitasking, managing distributed teams.

These most mentioned skills fall nicely into two of the conceptual skill areas in Henschel and Melonçon's matrix—*rhetorical proficiency* and *social proficiency*—and arguably in what might be considered additional conceptual skill areas (in a modified or expanded matrix): *technological literacy*, *communication*, and *management*. Henschel and Melonçon located technological literacy in the skill area of experimentation, and communication and management in the skill area of social proficiency. We return to the idea of a modified matrix later in Chapter 6, when we report needed competencies for content professionals. We want to note here, as well, the close alignment of these most mentioned skills (and their associated conceptual skill areas) with those skills identified by Hayhoe (2000); Rainey, Turner, and Dayton (2005); and Cleary (2021).

TABLE 5.2 Competency Studies Examining Job Advertisements or Practitioner Perspectives through Interviews or Surveys

Study	*Approach*	*Skills and Competencies*
Analyses of Technical Communication Job Ads		
Brumberger and Lauer (2015), "The Evolution of Technical Communication: An Analysis of Industry Job Postings"	Analyzed close to 1,000 US TC job postings from Monster.com. Queried a wide range of job titles. Examined three questions, one of which focused on competencies: • What professional competencies and personal characteristics are essential for success in the technical communication job market?	Professional competencies required: • Written communication • Project management • Editing • Visual communication • Research Personal characteristics desired: • Analytical/critical thinking • Collaboration • Detail oriented • Time management/ deadlines • Independence/initiative
Stanton (2017), "Do Technical/ Professional Writing (TPW) Programs Offer What Students Need for Their Start in the Workplace? A Comparison of Requirements in Program Curricula and Job Ads in Industry"	Examined 60 job ads posted to three different job boards to identify the skills and experiences required for entry-level technical writing positions. Queried only for "technical writer."	Skills mentioned most often: • Communication • General Software knowledge • Multitasking • Collaboration skills • Subject-matter expertise • Interpersonal skills • Technical writing skills • Detail oriented • Organizational skills Technical knowledge/ experience not listed as skill but discussed as a category. Subcategories included computer language knowledge/experience, markup language knowledge/experience, subject-matter knowledge, and foreign language knowledge.

(*Continued*)

TABLE 5.2 (Continued)

Study	*Approach*	*Skills and Competencies*
Kimball (2015), "Training and Education: Technical Communication Managers Speak Out"	Used modified Delphi method to assess the perceptions and attitudes of eight technical communication managers. The Delphi method included two rounds of surveys and focus groups, with each round "building on the results of the last"; "results of each round are shared with the participants" (p. 91). Relevant to skills and competencies, the main question asked: • What skills and competencies should we be teaching given desired credentials?	Most desirable skills (ranked in order): • Content development • Writing • Working in teams • Critical thinking • Audience analysis • Communication strategy • Information design • Information architecture • Document design • Managing distributed work Strategic skills valued: • Business skills • Content-area skills • Subject-matter knowledge • Iterative project management skills (e.g., Agile scrum competency) Note: End of round 2, four respondents advised programs to develop in students: • "systems thinking" skills • "interpersonal skills" such as "assertiveness and independence" • technical skills or domain knowledge in other professions, such as engineering or project management
Stanton (2017), Do Technical/ Professional Writing (TPW) Programs Offer What Students Need for Their Start in the Workplace? A Comparison of Requirements in Program Curricula and Job Ads in Industry	Surveyed five professionals who regularly hire entry-level technical writers. Asked, "What are the top 3–5 skills you seek in any candidate?" (This could be technical skills, software skills, or soft skills.)	Skills (top 3–5 sought in job candidates): • Technical writing • Oral communication • Technical Acumen/ knowledge • Editing • Interviewing • Collaboration • HTML/programming languages

(*Continued*)

TABLE 5.2 (Continued)

Study	*Approach*	*Skills and Competencies*
	Also asked, "Which soft skills they look for in a candidate?"	Soft skills mentioned as most important: • Motivation to help others • Strong interviewing skills • Independence • Interpersonal/ communication skills • Adaptability • Problem-solving skills • Teamwork
Shalamova, Rice-Bailey, and Wikoff, (2018), "Evolving Skill Sets and Job Pathways of Technical Communicators"	Surveyed 64 working technical communicators to identify their various job titles and the skills required for their positions. Also asked questions on what additional training or education they might pursue to improve skills.	Skills Deemed Important (ranked): • Writing • Interpersonal communication • Project management • Cross-functional teams • Teamwork • Presenting • Content management • Usability • Software • UX design • Distributed teamwork
Moore and Earnshaw (2020), "How to Better Prepare Technical Communication Students in an Evolving Field: Perspectives from Academic Program Directors, Practitioners, and Recent Graduates"	Interviewed 15 technical communicators in three different categories (four academic program directors, six practitioners, and five recent graduates). Among others, asked questions concerning how well-prepared graduates are for entering the profession.	Skills mentioned most often in interviews with the six practitioners: • Fundamentals of writing • Technical skills (ability to use different software programs/coding skills) • Technological literacy

Among the six studies, soft skills most mentioned included critical and analytical thinking, problem-solving, and independence and initiative. Technical communication managers in Miles Kimball's (2015) study also emphasized strategic skills of growing importance; these included business skills, content-area skills, subject-matter knowledge, and iterative project management skills—all skills that align with those that the industry leaders and hiring managers whom we interviewed identified as important.

These strategic skills, argued Kimball, are increasingly necessary for career advancement, as basic skills alone limit career paths.

While technical writing and written communication skills continue to be the most desired skills across diverse technical communication work contexts, it is unclear from these studies whether the jobs advertisements analyzed or participants interviewed or surveyed equated strong technical writing skills with strong structured authoring skills (that is, whether or not for some organizations and some participants "technical writing" meant "structured authoring"). Brumberger and Lauer (2015) did conclude that their data supported "the idea that content—as opposed to documents—drives the newer types of technical communication positions" and that the "emphasis on content necessitates a reexamination of academic curricula" (p. 239).

Competencies Connected to the Discipline of Content

We have seen numerous calls for curricular innovation in the content-focused facet of technical communication curriculum (e.g., Bridgeford, 2020a; Getto & Flanagan, 2024; Gu & Pullman, 2008; Swarts, 2024), and there is a growing body of scholarship that describes different approaches to teaching content design and particularly content strategy (e.g., Tham, Howard, & Verhulsdonck, 2022; Getto, 2018; Getto, Labriola, & Ruszkiewicz, 2020). However, there is a lack of empirical studies that specifically examine needed skills and competencies for work in the disciplinary areas within the content discipline. Tracy Bridgeford's (2020b) edited collection on *Teaching Content Management in Technical and Professional Communication* offered several chapters on teaching various activities within the context of content strategy and topic-based authoring. So too did Getto, Labriola, and Ruszkiewicz's (2020) edited collection on *Content Strategy in Technical Communication,* as well as Guiseppe Getto and Susan Flanagan's (2024) special issue of *Technical Communication* on "The How-to of Content Strategy: Teaching, Training, and Application." A missing component in these publications, though, was a close examination of the skills and competencies that practitioner leaders identify as necessary for entry- and advanced-level content-focused positions within organizations that have adopted structured authoring workflows.

To date, the approaches taken to identifying skills and competencies for content professionals have been syntheses of the content-focused scholarly and trade discourse (Batova & Andersen, 2017; Carliner, 2020), phenomenological research and personal conversations (Andersen, 2014), and experience- and research-informed views of the type of computational thinking required to understand and perform well in modern content ecosystems (Evia, 2018; Evia, Sharp, & Pérez-Quiñones, 2015). Our direct work in some of these approaches has zoomed our attention into higher order skills well aligned with those that Henschel and Melonçon (2014) identified in their explanatory layered literacies matrix.

Rebekka and our colleague Tatiana Batova conducted the first systematic literature review of scholarly and trade publications that discussed roles and skill sets needed for content-focused work in technical communication (Batova & Andersen, 2017). A key question guiding their review was "What skills do technical communicators need for career longevity in CCM [component content management] environments?" (p. 174). Their identified essential skills included the following for both entry- and advanced-level roles:

- Most important: *minimalist, structured writing*
- Additionally important:
  - *project management and leadership skills*
  - *collaboration skills* (ability to collaborate across teams and organizational units)
  - *content marketing skills* (understanding how content adds value for customers and how to measure that value)
  - *technology skills* (proficiency with markup languages and metadata, understanding of and ability to assess content technologies)
  - *global workflow skills* (e.g., developing processes that support translation and localization).

These skills, Batova and Andersen argued, are essential for content authors, technology specialists, content strategists, information architects, and other key roles in content ecosystems, as those serving in these roles are all "tasked with managing content and projects, contributing to cross-functional teams, and demonstrating the value they add to their organizations" (p. 185).

Like Batova and Andersen, Saul Carliner (2020) offered a synthesis of competencies needed for performing content-focused work. As a contributing author to Bridgeford's (2020b) edited collection *Teaching Content Management*, he reviewed chapters in the collection with the goal of expanding educators' understanding of the competency *CM*, one of the nine core competencies in the STC Competency Model (notably, in Chapter 2, we discuss how CM has also been used as an umbrella term in the academic discourse to refer to all things content). Carliner noted that CM in the STC Competency Model was limited by its focus on the use of websites. His expanded articulation of this competence included the ability to do the following:

- "describe the capabilities of content management software"
- "compose material for dynamic publication in a variety of media"

- "prepare a strategy for communicating a body of content employing content management technologies"
- "implement a content management system that addresses" defined publishing needs and work processes
- "use [data] reports to strengthen internal work processes and expand the reach [and usefulness] of the content" (p. 47).

These abilities begin to provide insight into what the conceptual skills of technological literacy, rhetorical proficiency, social proficiency, experimentation, abstraction, and critical system thinking look like when applied to content-focused work environments. Carliner did not distinguish between entry-level and advanced-level competencies but rather asserted that competence in CM was needed regardless of which of the four broad technical communication job roles one assumed in these environments: planner/designer role, creator role, production role, and manager role. These roles, as Carliner defined them (see pp. 48–52), map well, respectively, to the disciplinary areas of content strategy, content design, content engineering, and content operations and essential activities fused in those areas.

Drawing on findings from her phenomenological study of the CM discourse, Rebekka (Andersen, 2014) concluded that content professionals "who perform symbolic-analytic work in CM contexts draw on their *business knowledge*, *technical aptitude*, and *rhetorical expertise* [emphasis added] to solve complex information problems" (p. 143). These competency areas require skills in topic-based authoring, information design and information architecture, user-centered design and user-experience design, multimedia integrations and production, collaboration, business analysis, and management (p. 144). In her study that examined changes in the content industry and how those changes were affecting rhetorical and symbolic-analytic work in content-focused contexts, Rebekka argued that these skills "are essential for performing rhetorical work that includes analyzing, designing, authoring, and managing an organization's content corpus" (p. 143). Content professionals who wish to move into leadership positions—that is, more advanced positions as analysts, architects, strategists, and managers of content teams—need to develop these abilities to position themselves as orchestrators, coordinators, and negotiators of content tasks and practices (p. 143).

A key takeaway from Rebekka's study was an understanding that symbolic-analytic work skills were required for performing well in the disciplinary areas of content strategy, content engineering, and content operations (and they were also required for moving from entry-level roles in content design to more advanced roles in all the disciplinary areas of content):

> Those contributing to the planning, design, and implementation of the [CM] system are symbolic-analytic workers who apply their knowledge of rhetoric, business, and technology to complex information problems. They use skills in abstraction, systems thinking, experimentation, and collaboration to carry out the work of content strategy, which includes analyzing user and business needs, developing an information architecture for the organization's content corpus, creating unified processes, and developing a technology and change management strategy. These high-level, problem-solving activities take place in the content strategy framework whereas content authoring activities take place within the CM system, which is governed by the framework.
>
> *(Andersen, 2014, p. 138)*

In her study, Rebekka used the language of the CM system, which has evolved into what we refer to in this book as the strategies, processes, and technologies supporting digital content transformations. She used the language of "the content strategy framework" to refer to the strategic plan and execution of that plan for all things content—the plan defines and governs the CM system. Those who contribute to this framework are symbolic-analytic workers who are skilled in content strategy and, when examining a content corpus, can anticipate and model "all possible audiences, assemblages, displays, and use contexts of content topics" (p. 139). This ability is, at its core, what abstraction and critical system thinking look like in the content discipline.

More specifically, Rebekka described examples of what skills in abstraction, systems thinking, experimentation, and collaboration look like in the facilitation of content work.

- *Abstraction*—analyzing "user and business needs (particularly to audit and analyze the organization's content corpus), mak[ing] a business case for moving an organization to CM, and develop[ing] an information architecture to support the CM system" (p. 139).
- *Systems thinking*—understanding "the whole of an organization's content corpus and content needs...before they can begin developing an information architecture that requires them to define relationships among content topics and elements.... also understand[ing] the whole of an organization's business processes and defin[ing] unified content processes that align with business needs" (p. 140).
- *Experimentation*—experimenting "with new content technologies, conducting test runs and exploring configuration and customization

options"; bouncing "ideas off other stakeholders, gathering their input and refining ideas based on that input" (p. 140).

- *Collaboration*—working effectively on cross-disciplinary teams and with representatives from cross-organizational departments such as marketing, training, and engineering (pp. 140–141).

These higher order conceptual skills are required for anyone contributing to strategic decisions that define and govern CM systems, but they are also required for understanding, at a foundational level, how digital content transformations, and particularly structured authoring workflows, work—how the parts are embedded in layers of abstraction that, when combined, allow for both human and machine creation of content that can be published or server to many different outputs, audiences, and contexts.

Carlos, in his book *Creating Intelligent Content with Lightweight DITA*, articulated the concept of layers of abstraction in terms of the conceptual skill of *computational thinking* (Evia, 2018). He described how the concept of abstraction has been used in computing—that is, computing professionals and scholars approach a problem in layers:

> Abstraction allows humans to separate the 'layers' of a particular problem and work on each one individually without concern for the others. Then, as the layers are recombined, they work together to solve the problem in a process that can be represented as a replicable algorithm.
> *(p. 150)*

Just like computing professionals approach problems in layers of abstraction, content authors need to approach content in layers of abstraction. A layer of abstraction approach, what Carlos associated with computational thinking, encourages authors to "view content as non-linear and envision more than one outcome based on principles of reuse and single-sourcing" (p. 151). Some elements within a topic of raw content are viewed as variables that change while other elements are viewed as requirements that do not change; all elements follow a particular order defined by a schema, but different elements are included in different outputs based on defined user needs and contexts of use. These are the kinds of layers of abstraction, or computational thinking, that intelligent content creation requires.

Computational thinking, then, as a conceptual skill, is a way of thinking "before structuring information and seeing its future applications"; it is a way of thinking about possibilities for final outcomes, about seeing "what the content can accomplish once it is liberated from ... word processing or web presentational models" (p. 152). Carlos emphasized the role of human abstraction in the automation of content development, management, and delivery processes—"machine automation," he argued, "cannot happen without human abstraction." Automation routines for content, as a key goal of digital content transformations, can happen only after "authors (and not necessarily programmers or developers)" "get involved in a structured process involving layers of abstraction and replicable rules" (p. 153). Abstraction as a skill that encompasses computational thinking, then, is essential for performing content-focused work in organizations that have adopted digital content transformations. While not a skill that surfaced in analyses of job advertisements or reports of interviews or surveys, or that surfaced in Batova and Andersen's (2017) or Carliner's (2020) syntheses of skills for content-focused work, abstraction should hold an important place in any competencies framework for preparing content professionals.

The Need for Industry Perspectives That Inform Competency and Curriculum Discussions

Keeping abreast of technical content trends and making sense of what they mean for future content professionals is proving to be an increasingly difficult task, for both technical communication researchers and educators. The researchers who have taken up this task have produced a small, but growing body of literature that informs our collective understanding of ways in which professional practice in the content-focused facet of the field is changing and how we as educators might innovate curricula and pedagogical approaches in response (see Batova & Andersen, 2017; Bridgeford, 2020b; Evia, 2018; Getto, Labriola, & Ruszkiewicz, 2020; Swarts, 2020, 2024; Tham, Howard, & Verhulsdonck, 2022; Verhulsdonck, Howard, & Tham, 2021). As we note in Chapter 1, these publications contribute important articulations of what are now well-established practices such as structured authoring and content strategy, and they offer both theoretical frameworks for which to understand these practices and specific recommendations for revising existing or developing new curricular components, from student learning outcomes to courses to assignments.

What is not represented in the literature specifically focused on technical content to date, and what Chapter 6 includes, are perspectives from industry leaders and recent alumni on the competencies and qualities that graduates need to both gain access to and then succeed and advance in entry-level content-focused positions. The literature has yet to capture perspectives, as well, on curricular innovations that would prepare this next generation of content professionals for leading and innovating in the content profession beyond entry-level positions. These perspectives are imperative to understanding how we as educators can better align our programs and curricula with practices that now define the discipline of content, and particularly with practices that now define the disciplinary area of content design, where we locate the activities of creating in-product and technical content. While research projects aimed at understanding industry perspectives such as those of Carnegie and Crane (2018), Duin and Tham (2018), and Shalamova, Rice-Bailey, and Wikoff (2018) serve as models for how administrators can better align curricula with local industry needs, we focus in the next chapter on content practices specifically (as opposed to technical communication practices generally) and on perspectives from leaders working across industries in varied localities, most of whom had 25+ years of experience.

Note

1 Descriptions are drawn from a 2016 special issue of *Intercom* on the certification program and a 2017 special issue of *Intercom* on "Tech Comm's Core Competencies."

References

Agnew, B. (2017, March). Content management. *Intercom*, 21–23.

Andersen, R. (2014). Rhetorical work in the age of content management: Implications for the field of technical communication. *Journal of Business and Technical Communication*, *28*(2), 115–157.

Andersen, R., & Evia, C. (Eds.). (2018). Perspectives on preparing technical communication professionals for today and the future [Special issue]. *Communication Design Quarterly*, *6*(3).

Baehr, C. (2016, January). The foundation exam and its nine areas of competency. *Intercom*, 10–11.

Baehr, C. (2017, March). Organizational design. *Intercom*, 11–12.

Batova, T., & Andersen, R. (2017). A systematic literature review of changes in roles/skills in component content management environments and implications for education. *Technical Communication Quarterly*, 26(2), 173–200.

Blakeslee, A. M. (2001). Bridging the workplace and the academy: Teaching professional genres through classroom-workplace collaborations. *Technical Communication Quarterly*, *10*(2), 169–192.

Bridgeford, T. (2020a). Content management: A pedagogical exigency. In T. Bridgeford (Ed.), *Teaching content management in technical and professional communication* (pp. 1–15). New York: Routledge.

Bridgeford, T. (Ed.). (2020b). *Teaching content management in technical and professional communication*. New York: Routledge.

Brown-Hoekstra, K. (2017). Reviewing and editing. *Intercom*, 18–20.

Brumberger, E., & Lauer, C. (2015). The evolution of technical communication: An analysis of industry job postings. *Technical Communication*, *62*(4), 224–243.

Brumberger, E., Lauer, C., & Northcut, K. (2013). Technological literacy in the visual communication classroom: Reconciling principles and practice for the 'whole' communicator, *Programmatic Perspectives*, *5*, 3–28.

Cargile Cook, K. (2002). Layered literacies: A theoretical frame for technical communication pedagogy. *Technical Communication Quarterly*, *11*(1), 5–29.

Carliner, S. (2020). Content management: Preparing technical communication students for the realities of the workplace. In T. Bridgeford (Ed.), *Teaching content management in technical and professional communication* (pp. 40–56). New York: Routledge.
Carliner, S., & Chen, Y. (2018). Who technical communicators are: A summary of demographics, backgrounds, and employment. *Intercom*, 8–12.
Carnegie, T., & Crane, K. (2018). Responsive curriculum change: Going beyond occupation demands. *Communication Design Quarterly*, *6*(3), 25–31.
Carter, M. (2007). Ways of knowing, doing, and writing in the disciplines. *College Composition and Communication*, *58*(3), 385–418.
Cleary, Y. (2021). *The profession and practice of technical communication*. New York: Routledge.
Dicks, S. (2002). Cultural impediments to understanding: Are they surmountable? In B. Mirel & R. Spilka (Eds.), *Reshaping technical communication: New directions and challenges for the 21st century* (pp. 13–25). Mahwah, NJ: Erlbaum.
Duin, A. H., & Tham, J. (2018). Cultivating code literacy: Course redesign through advisory board engagement. *Communication Design Quarterly Review*, *6*(3), 44–58.
Durack, K. (2003). Instructions as "Inventions": When the patent meets the prose. In T. Kynell-Hunt & G. Savage (eds.), *Power and Legitimacy in Technical Communication, Volume I: The Historical and Contemporary Struggle for Professional Status* (pp. 15–37). Amityville, NY: Baywood.
Evia, C. (2018). *Creating intelligent content with Lightweight DITA*. New York: Routledge.
Evia, C., & Priestley, M. (2016). Structured authoring without XML: Evaluating Lightweight DITA for technical documentation. *Technical Communication*, *63*(1), 23–37.
Evia, C., Sharp, M., & Pérez-Quiñones, A. (2015). Teaching structured authoring and DITA through rhetorical and computational thinking. *IEEE Transactions on Professional Communication*, *58*(3), 328–343.
Getto, G. (2018). A prototype theory for content strategy education: Training content strategists within the academy. In *Proceedings of the 36th ACM International Conference on the Design of Communication* (pp. 1–6).
Getto, G., & Flanagan, S. (Eds.). (2024). The how-to of content strategy: Teaching, training, and application [Special issue]. *Technical Communication*, *71*(3).
Getto, G., Labriola, J., & Ruszkiewicz, S. (2020). *Content strategy in technical communication*. New York: Routledge.
Giammona, B. (2004). The future of technical communication: How innovation, technology, information management, and other forces are shaping the future of the profession. *Technical Communication*, *51*(3), 349–366.
Gillenwater, J. (2017). Visual communication. *Intercom*, 15–17.
Gu, B., & Pullman, G. (2008). Mapping out the key parameters of content management. In G. Pullman & B. Gu (Eds.), *Content management: Bridging the gap between theory and practice* (pp. 1–12). Amityville, NY: Baywood.
Hart-Davidson, W. (2001). On writing, technical communication, and information technology: The core competencies of technical communication. *Technical Communication*, *48*(2), 145–155.

Hayhoe, G. (2000). What do technical communicators need to know? *Technical Communication*, *47*(2), 151–153.

Henning, T., & Bemer, A. (2016). Reconsidering power and legitimacy in technical communication: A case for enlarging the definition of technical communicator. *Journal of Technical Writing and Communication*, *46*(3), 311–341.

Henschel, S., & Melonçon, L. (2014). Of horseman and layered literacies: Assessment instruments for aligning technical and professional communication undergraduate curricula with professional expectations. *Programmatic Perspectives*, *6*(1), 3–26.

Houser, A. (2017a, March). Content development. *Intercom*, 1–10.

Houser, A. (2017b, March). Written communication. *Intercom*, 13–14.

Hovde, M. R. (2022). Procedural knowledge and discourse in technical communication: Easy as 1, 2, 3? In J. Schreiber & L. Melonçon (Eds.), *Assembling critical components: A framework for sustaining technical and professional communication* (pp. 137–168). WAC Clearinghouse: University Press of Colorado.

Hovde, M. R., & Renguette, C. C. (2017). Technological literacy: A framework for teaching technical communication software tools. *Technical Communication Quarterly*, *26*(4), 395–411.

International Technology Education Association. (2006). *Technological literacy for all: A rationale and structure for the study of technology*. Reston, VA: Author.

Johnson, R. (1998). *User-centered technology: A rhetorical theory for computers and other mundane artifacts*. Albany, NY: SUNY Press.

Johnson-Eilola, J. (1996). Relocating the value of work: Technical communication in a post-industrial age. *Technical Communication Quarterly*, *5*(3), 245–270.

Kimball, M. (2015). Training and education: Technical communication managers speak out. *Technical Communication*, *62*(2), 135–145.

Lanier, C. (2009). Analysis of the skills called for by technical communication employers in recruitment postings. *Technical Communication*, *56*(1), 51–61.

Littlejohn, A., Beetham, H., & McGill, L. (2012). Learning at the digital frontier: A review of digital literacies in theory and practice. *Journal of Computer Assisted Learning*, *28*(6), 547–556.

Melonçon, L., & Schreiber, J. (2022). Introduction: Promoting a sustainable collective identity for technical and professional communication. In J. Schreiber & L. Melonçon (Eds.), *Assembling critical components: A framework for sustaining technical and professional communication* (pp. 3–16). WAC Clearinghouse: University Press of Colorado.

Moore, L. E., & Earnshaw, Y. (2020). How to better prepare technical communication students in an evolving field: Perspectives from academic program directors, practitioners, and recent graduates. *Technical Communication*, *67*(1), 63–82.

Newmark, J., & Bartolotta, J. (2021). Creating the "through-line" by engaging industry certification standards in SLO redesign for a core curriculum technical writing course. In M. Klein (Ed.), *Effective teaching of technical communication theory, practice, and application* (pp. 147–165). WAC Clearinghouse: University Press of Colorado.

Pohland, L., & Lyons, C. (2016, January). Certification and STC's certified professional technical communicator. *Intercom*, 6–9.

Rainey, K., Turner, R., & Dayton, D. (2005). Do curricula correspond to managerial expectations? Core competencies for technical communicators. *Technical Communication*, *52*(3), 323–352.

Reich, R. (1992). *The work of nations*. New York: Vintage Books.

Rude, C. D. (2015). Building identity and community through research. *Journal of Technical Writing and Communication*, *45*(4), 366–380.

Selber, S. (2004). *Multimodal literacies for a digital age*. Carbonate, IL: Southern Illinois University Press.

Shalamova, N., Rice-Bailey, T., & Wikoff, K. (2018). Evolving skill sets and job pathways of technical communicators. *Communication Design Quarterly*, *6*(3), 14–24.

Slattery, S. (2005). Technical writing as textual coordination: An argument for the value of writers' skill with information technology. *Technical Communication*, *52*(3), 353–360.

Stanton, R. (2017). Do technical/professional writing (TPW) programs offer what students need for their start in the workplace? A comparison of requirements in program curricula and job ads in industry. *Technical Communication*, *64*(3), 223–236.

Swarts, J. (2020). Writing about structure and DITA. In T. Bridgeford (Ed.), *Teaching content management in technical and professional communication* (pp. 155–175). New York: Routledge.

Swarts, J. (2024). Afterword. In C. Evia (Ed.), *Content operations from start to scale: Perspectives from industry experts* (pp. 155–162). Blacksburg: Virginia Tech Publishing.

Tham, J., Howard, T., & Verhulsdonck, G. (2022). Extending design thinking, content strategy, and artificial intelligence into technical communication and user experience design programs: Further pedagogical implications. *Journal of Technical Writing and Communication*, *52*(4), 428–459.

Verhulsdonck, G., Howard, T., & Tham, J. (2021). Investigating the impact of design thinking, content strategy, and artificial intelligence: A "Streams" approach for technical communication and user experience. *Journal of Technical Writing and Communication*, *51*(4), 468–492.

Whiteside, A. (2003). The skills that technical communicators need: An investigation of technical communication graduates, managers, and curricula. *Journal of Technical Writing and Communication*, *33*(4), 303–318.

Wilson, G. (2001). Technical communication and late capitalism: Considering a postmodern technical communication pedagogy. *Journal of Business and Technical Communication*, *15*(1), 72–99.

6

A COMPETENCIES FRAMEWORK FOR TECHNICAL CONTENT PROFESSIONALS

Cultivating Critical Pragmatic Practitioners in the Discipline of Content

How to prepare the next generation of content professionals has been the driving question behind this research project. Given what we had learned from our conversations with industry colleagues and from the many conference sessions we had attended over the years focused on workforce and staffing needs, we knew that educators in the field had much work to do if they wanted graduates of their programs to be positioned to take on content-focused roles that afforded them meaningful agency in their future organizations. Getting in the door—landing that first entry-level position—was key.

As we argue in Chapter 1, much is at stake if humanities-trained content professionals are not foundational contributors to the design and iterative improvement of content initiatives. Humanities-trained content professionals can bring much-needed human-centered design values and critical sensibilities to practices essential to digital content transformations, such as minimalist writing, AI content generation, taxonomy development, and content modeling.

We very much agree with researchers and educators in the field who have argued for preparing students to be what Lisa Melonçon and Joanna Schreiber (2022) call "critical pragmatic practitioners," or "students (and future professionals) who can reflect and critique and who can also make, produce, and communicate as a key result of academic degrees" (p. 24).

DOI: 10.4324/9781003451730-10

Questions along these lines that we have kept at the forefront of our research include the following:

- What does a critical pragmatic practitioner in the content discipline look like?
- What skills and qualities does a student need to gain access to an entry-level position in the discipline and to advance into more strategic positions that afford them meaningful agency and opportunities to lead and innovate?
- What skills and qualities does a student need to gain access to an entry-level position in the discipline and to advance into more strategic positions that afford them opportunities to advocate for marginalized and underrepresented users and team members and socially just systems and practices?

In this chapter, we situate within the literacy frameworks discussed in Chapter 5 the skills, competencies, and qualities that our research participants identified as necessary for performing knowledge work in the discipline of content. We specifically expand Henschel and Melonçon's (2014) explanatory matrix of conceptual and practical skills to map with the competencies entry-level content professionals need to not only land content-focused positions but also meaningfully contribute to the content initiatives of their organizations—as critical pragmatic practitioners. In expanding the matrix, we introduce a competencies framework for content professionals that we hope administrators and educators will find useful for the purpose of planning program and curricular innovations. We end with a discussion of curricular gaps and opportunities for innovation in US undergraduate programs in technical communication.

Skills and Competencies for Content Professionals—What We Learned from Industry and Academic Leaders

In our interviews with industry leaders working in one or more disciplinary areas related to technical content and with academic leaders working in undergraduate programs that continually update curriculum to align with industry practices, we asked several questions about what they saw as implications of changing activities, roles, and hiring needs for educational programs, particularly undergraduate degree programs, that aim to prepare future content professionals. Specifically, we asked about the most important skills and competencies that students need to learn and about how undergraduate programs can best prepare students to become leaders and innovators in the content-focused sector of technical communication. In our survey of hiring managers, we asked just one question focused on education—and that was about what educators can do to best prepare students to be leaders and innovators.

Recent alumni of undergraduate programs who were working in full-time content-focused positions had much to say about their educational experiences and offered recommendations for programs as well. We asked those whom we interviewed to speak to the extent to which their college experience had prepared them for their current positions and for more strategic positions, and to what they wished they had learned during their undergraduate education. We include a summary of alumni reflections on their undergraduate training in an Alumni Perspectives breakout box (refer to Box 6.1).

Table 6.1 presents the questions related to implications for education that we asked of these groups.

TABLE 6.1 Survey and Interview Questions Focused on Implications of Changing Activities, Roles, and Hiring Needs for Technical Communication Education

Survey and Interview Questions	
Interviews—Industry Leaders	• What do you see as implications of changing roles and hiring needs for education in technical communication (TC)—including undergraduate degree programs, minors, and certificates? • What do you see as the most important skills and competencies for students completing degrees in TC to learn? • How can we best prepare students to become leaders and innovators in the profession? What does it mean to be a leader and innovator in the profession? • What advice would you give administrators of TC programs to make sure their programs were meeting the needs of the profession?
Interviews—Academic Leaders	• What do you see as implications of changing roles and hiring needs for education in TC (undergraduate degree programs, minors, certificates)?

(*Continued*)

TABLE 6.1 (Continued)

	• What do you see as the most important skills and competencies for students completing degrees in TC to learn? • How can we best prepare students to be leaders and innovators in the profession? What does it mean to be a leader and innovator in the profession? • What skills and competencies does your program teach/emphasize? Why? And how?
Survey—Managers	• What is one thing educators can do to best prepare students to be leaders and innovators in the technical communication profession?
Interviews—Alumni	• To what extent do you feel your undergraduate training prepared you for your current position? What courses did you find most helpful? Least? • What do you wish you had studied/learned during your undergraduate training? Why? • To what extent do you feel your undergraduate training prepared you to contribute to more strategic TC activities? To innovate on your team or within the larger organization? • In your opinion, what would an ideal TC undergraduate curriculum look like?

This section reports on what we learned from industry and academic leaders, hiring managers, and alumni. We share their perspectives on the most important skills and competencies that students completing degree programs in technical communication should learn, as well as their perspectives on how educators can best prepare students to be leaders and innovators. As we mentioned in Chapter 5, no empirical study to date has specifically examined needed skills and competencies for work in the disciplinary areas of content.

In response to questions about implications for education in technical communication, both industry and academic leaders offered advice for educators on how they could ensure their programs were meeting the needs of the profession. In Chapter 7, we share these recommendations for curriculum and program design, which we articulate in terms of enhancing curricular pathways.

In Chapter 4, we identified and elaborated on the four skills and qualities that leaders and hiring managers in the industry side of technical content most mentioned as needed for entry-level content-focused jobs. These included the following:

- Technical knowledge and skills
- Learning mindset
- Structured authoring
- Writing and communication skills.

Not surprisingly, the skills and competencies most mentioned in response to our questions about educating future content professionals aligned with those most mentioned in response to questions about skills needed for technical content-focused positions. These skills and competencies included the following, in order of importance based on our data analysis results:

1 Structured authoring
2 Systems knowledge and thinking
3 Technical knowledge
4 Communication skills
5 Collaboration mindset
6 Learning mindset and initiative.

Structured authoring was discussed twice as often as any other competency area for students completing degree programs aimed to prepare them for work as content professionals. This has become the key foundational practical skill for entry-level positions within the disciplinary area of content design. Students need to understand the basic concepts of structured authoring, such as modular content, typed content, semantic content, reuse, and the separation of content from form, before they can begin to understand the bigger picture of a content ecosystem or a digital content transformation initiative. Systems knowledge and thinking are about understanding this bigger picture and how people, processes, and technologies need to interrelate and function within a larger system to enable content to be treated as a valuable, computable asset.

Technical knowledge, communication skills, and a collaboration mindset are essential conceptual skills that encompass more practical skills, such as proficiency with markup languages, audience analysis, and interpersonal communication. These essential higher order skills are about ways of thinking and doing, working with others, and advocating for change and audiences. They are importantly connected with mobility—that is, with moving into more advanced content-focused positions. While not so much a skill as a quality per se, having a learning mindset and a drive to initiate ideas and solve problems was identified as a key quality that educational programs need to cultivate throughout curricula. This quality is about having the mindset that lifelong learning, adaptability, and curiosity are central to what it means to be an impactful professional in the content discipline. It is about learning how to be flexible and embracing the fact that continually learning new technology, processes, and approaches is just part of what it means to be a content professional. Importantly, in the discipline of content, the role of educational programs is to help students build foundational knowledge, skills, and mindsets; as content professionals, they will need to engage in other educational experiences to build on those foundations as well as the skills that they develop through situated practice.

When we specifically asked about educating future leaders and innovators in the discipline of content—that is, about how educators can prepare students to perform well beyond the entry-level position—leaders and hiring managers talked mostly about students needing to develop systems knowledge and thinking, collaboration skills (particularly cross-team collaboration and cross-functional leadership), and communication skills (notably, a learning mindset and initiative cut across responses).

What they did not talk about was structured authoring, and few mentioned technical knowledge, both of which are foundational skills for landing entry-level content-focused positions. These skills are assumed as foundational for anyone moving into more strategic roles that afford opportunities to lead and innovate. What leaders did talk about extensively was the need for students to learn about business management, strategy, and processes, as well as product development processes; to learn and practice strategies for collaborating and advocating within and across teams and organizations; and to learn how to build research-informed business cases focused on process improvements. They talked, as well, about students needing to be strong, agile communicators and big picture, innovative thinkers who can identify problems and propose data-backed solutions.

These qualities and skills are necessary for moving from entry-level roles to more advanced roles as senior writers or managers or as content architects or strategists. While structured authoring experience and technical knowledge may be enough for students to land entry-level content-focused positions, these competencies alone will not afford students opportunities to move into more advanced roles. Because we as educators are interested in graduating students who not only land good jobs as content professionals but who are well-positioned to assert agency and have impact in their organizations, we need to help students develop competencies that will afford them opportunities to innovate and lead—to help shape the ways in which processes and technologies supporting digital transformation workflows impact both end-user experiences and work environments.

Certainly, students will learn a great deal on the job and gain valuable experience that will open doors to opportunities to innovate and lead, but the more we can prepare them for symbolic-analytic work in modern content ecosystems, the better equipped they will be for solving challenging problems and assuming leadership roles.

A point of discrepancy between what some of our colleagues in academia perceive as skills and our experience and interpretation of research findings is that we do not see structured authoring (and even standards for structuring and publishing content like Darwin Information Typing Architecture (DITA)) as a skill set to be fully learned on the job. Over a couple of decades, we have been teaching, researching, and collaborating in authoring and publishing environments and life cycles guided by structured authoring and component content management as methodologies and not specific skills or tools. We do see specific software tools and platforms, such as Oxygen XML, Heretto, or Adobe FrameMaker, as tools implementing those methodologies. These software tools, and not the concepts behind them, do belong in the realm of structured on-the-job training materials that should not be at the center of college curricula.

Importantly, in addition to questions about needed skills and competencies for entry-level roles and for moving into more advanced roles, we asked our interview and survey participants what it means to be a leader and innovator in organizations that have adopted digital content transformation workflows. In a nutshell, these participants generally agreed that being a leader and innovator means being someone who is "always innovating and looking to improve processes" and who is "always learning and adapting." Below are representative responses from survey participants to the question of what it means to be a leader and innovator in technical communication:

- Someone who finds "new and better ways to provide information to our users."
- "Whether this is in the production and publishing of material, or in the presentation and organization of information, a leader brings innovation into the team and makes it work for the better."
- Someone who is "a thought leader, good communicator, a person who finds solutions to daily issues faced by teams and adapts to new trends in technical communication within the organization."
- Someone who thinks of tomorrow, who thinks about "how content is changing and how it will [continue to] change and adapt with technology" and about how users' needs for content and interactions with content are also changing. Someone who "experiments and comes up with new ways to present content."

- It means being "malleable, ready to adapt newer processes based on demands of the product."
- "It means we are constantly learning and growing and developing new best practices that meet the demands of our future clients, not just our clients where they are today."
- "It means not just being an expert on internal products, processes, and systems but also being 'an expert in the profession' and keeping up with trends and proven practices."

In the sections that follow, we draw on our interview and survey findings to articulate what these qualities and skills—what we call "competencies" from here on—look like for content professionals at the entry and more advanced levels, and why these competencies should be a central focus of curricular pathways that prepare students for content-focused work. We also describe competencies for mobility—gleaned from questions we asked about how educators can prepare students to be leaders and innovators in the discipline of content. We include a breakout box, as well, that summarizes the perspectives of five alumni of technical communication undergraduate programs who were working as content professionals when they participated in our study (refer to Box 6.1).

BOX 6.1 ALUMNI PERSPECTIVES

Recent alumni of undergraduate programs in technical communication and related fields who were working full-time jobs related to content had much to say about their educational experiences. We asked those whom we interviewed to speak to the extent to which their undergraduate education had prepared them for their positions, what they wished they had learned during their undergraduate courses, and what they thought an ideal undergraduate curriculum should look like.

Here, we focus on the responses of Kat, Jessica, Holly, Nick, and Lang—five alumni who were working in technical content-focused jobs at the time of our interviews. Their job titles ranged from technical specialist to technical writer to content developer; three worked for software companies, one for a manufacturing company, and one for a content solutions consultancy. As undergraduates, all five took courses in a technical discipline and/or in structured authoring.

Their responses to our questions focused mostly on four themes: technical training, multidisciplinary approaches, business knowledge, and communication and writing skills.

Technical Training

Technical training included references to technical topics, markup languages, the DITA standard, and particular tools. The five alumni talked a lot about how much they valued their technical preparation and taking courses in computer science or engineering. Classes that Kat and Jessica found most valuable included web development (focusing on HTML and CSS) and user documentation (focusing on XML and DITA). Holly, too, valued a course that she took focused on DITA. Kat, who worked for a content solutions consultancy, lamented not having the opportunity to take a course in customizing XML publications with XSLT (Extensible Stylesheet Language Transformations). She emphasized the importance of undergraduate curricula focusing on "the componentizing of content" and processes for publishing "content in a technical way."

Nick and Lang pointed to courses in computer science as being particularly helpful in their positions at the time of our interviews. Nick valued the engineering perspective that programming and computer science courses offered for understanding software and the different technologies he used for creating both internal documentation and external customer-facing documentation for cloud-based environments. Lang, as someone who worked daily with engineers on upcoming software releases via Agile development workflows, wished she had been introduced to terms such as Agile, scrums, sprints, and processes for documenting a feature end-to-end as an undergraduate. She emphasized the need for technical communication curricula to be both writing intensive and focused on computing and data science.

Multidisciplinary Approaches

Multidisciplinary approaches as a theme focused on how the field of technical communication intersects with other fields such as engineering, and how knowledge of technical subjects supports communications with subject matter experts and informs documentation approaches. Lang, Jessica, and Nick were particularly thankful for their experiences as undergraduates working on teams with students from diverse academic majors. Lang shared, "Working in small groups of about five that included engineers was perfect"; Lang was able to contribute her expertise in writing reports while the engineers were able to contribute their expertise in math and data analysis. Whereas Lang highly valued having completed a BS that required her to take engineering courses, Jessica highly valued her interdisciplinary design minor and the option to take course electives outside of her department to fulfill requirements for her major in technical communication.

Business Knowledge

Business knowledge included references to leadership, project management, change management, and the ability to make business arguments. Jessica and Holly spoke mostly about the importance of understanding business processes and being able to manage large projects and advocate for both themselves and for content development teams. Holly, a content developer for a multinational technology company, was tasked with managing multiple internal and external customer-facing documentation projects at a time and participated in daily scrum meetings with support teams and development teams. She found the projects and meetings exhilarating, but she shared that she would have been better prepared for her position had she taken more leadership courses and business-oriented courses while she was an undergraduate student.

Jessica, a technical writer for a fluid handling equipment manufacturer that was, at the time of our interview, beginning to transition from a largely unstructured to a structured authoring environment, wanted to take on a meaningful role in the transition. She knew the migration to DITA would require skills in change management and project management. When reflecting back on her undergraduate education, she wished she had taken more courses focused on project management, learning how to build requirements documents, how to gain buy-in from stakeholders, how to manage change, and how to advocate for oneself, one's ideas, and larger changes in an organization. She considered these business skills as crucial to the success of the DITA migration.

Communication and Writing Skills

Kat, Jessica, Holly, Nick, and Lang all valued the emphasis on communication and writing skills in their undergraduate programs. Oral and interpersonal communication skills were critical for working closely with different teams, particularly engineering teams, and for presenting ideas and project progress persuasively in meetings, such as daily scrum meetings. Holly mentioned being surprised by how often she was required to present to large groups and shared that she wished she had taken more courses in business and public speaking as an undergraduate. The five alumni stressed the importance of strong writing skills in their positions and were grateful for having taken courses as undergraduates that helped them develop these skills. The ability to tailor content to specific audiences, apply plain language strategies to writing, and effectively peer review other writers' work were skills that the alumni used daily.

Structured Authoring

As we reported in Chapter 4, the ability to create topic-based, well-structured content is widely considered a foundational skill among hiring managers for content-focused jobs. Managers expect job candidates, at a minimum, to have a conceptual understanding of structured authoring as a methodology and to be familiar with the technologies and standards available for supporting structured authoring.

Regardless of the standard or language used (e.g., DITA, JSON, or even Markdown), structured authoring is a modular approach to writing content that can be reused in multiple contexts and published in different formats. It is a methodology for writing structured content, which, as we argue in Chapter 3, is foundational for digital transformations. Structured content is the building block for transformations, offering both its producers and consumers vast affordances:

> Modules can be seamlessly integrated and quickly assembled into large, complex documents; they can be simultaneously delivered to chatbots, portal sites, mobile apps, and other channels; they can be dynamically delivered as personalized content experiences; they enable continuous content development and publishing; and they allow organizations to automate repetitive tasks (reduce "friction"), scale up content development and publishing quickly, and provide high quality source content to generative AI bots.
>
> *(Point reiterated here from Chapter 2, p. 69)*

Not surprisingly, structured authoring was the competency stressed most by industry and academic leaders in terms of competencies that educational programs need to cultivate in students who want to work as content professionals post-graduation. Other important competencies discussed—systems thinking and knowledge, technical proficiency, rhetorical agility and communication skills, and a collaboration mindset—connect back to an understanding that structured content is the foundation of digital content transformations.

The principal consultant and co-founder of an authoring solutions company had this to say about how structured authoring needs to be core in any curriculum preparing technical writers:

> We need to … really look at how we need to teach writing differently for technical writers. All right? We need to learn how to write for intent, we need to learn how to structure information and think about structure differently. There's far too many academic programs that are not talking about DITA, for example. And that still baffles my brain. I don't

> understand why academic programs in technical writing are not focusing on structured authoring. It doesn't have to be DITA, but DITA is probably the best place to start. And so we have to revisit how we treat or how we teach writing for technical writers.

Another industry leader, founder and principal of a content solutions consultancy, went so far as to suggest that educational programs "need to build structuralists and semanticists"—that is, "learners who can discern and communicate the structure of content and who understand the meaning of content objects and who can then work well with their counterparts in strategy and operations." In a more basic sense, building *structuralists* and *semanticists* means teaching structured authoring as a way of thinking and a process that includes activities such as conducting a content audit, analyzing content, typing or mapping content, defining structure using a schema or template, defining metadata and taxonomies to use for categorizing and tagging content elements and for enabling reuse, and applying principles of minimalism and plain language.

Structured authoring as a way of thinking, as articulated by one consultant, is about understanding "content as units, or genres of small units of information, that can be assembled in many ways to become larger genres"—it's about understanding the reuse and multiple-use potential of the same content units in different structures. It's also about understanding users and usage contexts. The same consultant continued,

> Who are these people we're talking to, what do they need, what questions are they asking? Because each question or each type of question comes with a requirement for a particular block of content as defined formally in the schema or the DTD.

This way of thinking about content as structured, small units of information that answer specific user questions is key to developing proficiency in structured authoring; so too is thinking of these units as separate from format, as only at the stage of publishing is format applied to the units via stylesheets. It is this separation that allows for content to be easily reused and updated, and it allows for multichannel publishing and personalized filtering for multiple audiences or products.

Developing competency in structured authoring means also learning how to write for reuse and write with an eye toward growth and translation. Two industry leaders spoke about writing for reuse as a required core skill for structured authoring, one that students need to develop:

> Because we become very, very sensitive to contextualization. How do we write things but place them in the places where they're gonna then

> be usable in a variety of ways? And then, ultimately, the fact that the content, at least by my definition, being potential information, means something is gonna happen to it. So [students] have to have a familiarity [with the big picture].
>
> —*Principal researcher and managing director of a content research consultancy (interviewee)*

> How do you write content that doesn't just follow AP style or this style guide or that style guide, but … [that] can appear in six different places? And how do you use your creativity to write this piece of content in a simple, easy to understand empathetic manner that can also exist with different phrases around it? I think that's a change in the way we write from how you're taught growing up.
>
> —*Senior director of content strategy for a web services subsidiary (interviewee)*

At the core of writing well for reuse is minimalism, which requires discarding information irrelevant to the immediate purpose of a unit of content, including discarding transitional elements. Many industry and academic leaders, when talking about structured authoring as a required competency, also emphasized the command of a publication's target language. Part of being a strong writer of structured content—being able to write for reuse and with an eye toward translation—is having a strong grasp of how to write clear and concise sentences—that is, having a strong grasp of how to employ principles of plain language to achieve clarity, accessibility, reuse, and translation goals.

Systems Knowledge and Thinking

In response to our question about the most important skills and competencies for aspiring content professionals completing undergraduate degrees to learn, systems knowledge and thinking emerged as a strong second theme—after structured authoring.

Industry leaders in particular emphasized the need for students to gain an understanding of the big picture—of the what, why, and how of digital content transformations and the business strategies and business and technical processes that support product and content life cycles. They spoke of the need for programs to prepare students for working in business environments, to focus more on business processes, Agile development processes, product development processes, and content workflows and life cycles. They also spoke of the need for students to understand how technical content teams are connected to other teams concerned with product and content development. One academic leader touched on this as well: When alumni of her program were asked the question, "What were you least prepared for?," they talked mostly about working in a business context.

A big-picture view—or systems view—of content (including diverse human authors and users, tools, and even AI models) is what we might call the new rhetorical situation that content professionals must examine (or at least take into account) when making design, strategy, engineering, or operations decisions at any level, from that of the entry-level technical writer to that of the director of content operations. In content ecosystems where rhetorical agency lies with the interdisciplinary team charged with designing and overseeing content ecosystems, as opposed to with the lone writer, content professionals must be prepared to work and think at a system level and to engage stakeholders from across an organization; so too must they be prepared to apply rhetorical and technical expertise as well as social proficiency to both understand complex products and systems and contribute meaningfully to content ecosystems. A systems view goes beyond the realm of writer-as-agent to the processes, people, and technologies that shape rhetorical agency.

For students, gaining a system-level view of the content discipline is important for understanding ways in which content design, content strategy, content engineering, and content operations relate. As Robert Reich (1992) described, a systems thinker aims to understand how different parts of the whole are related and can, when presented with a problem or issue, "discern larger causes, consequences, and relationships" (p. 30). This is not to say that students need advanced skills at system-level planning and engineering to land entry-level jobs; rather, the expectation is that they will gain these skills through work experience. But students who want to work as content professionals and who want opportunities to advance in their positions and move into new roles in other disciplinary areas of content need to develop systems knowledge and gain expertise in systems thinking. This conceptual skill—that is, systems knowledge and thinking—is one that both industry and academic leaders say should begin to be cultivated in educational programs that prepare students as content professionals.

Because content professionals increasingly work alongside engineers and developers in Agile development workflows that prioritize continuous product development, students need to gain familiarity with the concept of product life cycles. They need to understand how product development processes are connected to business processes, and they would do well to have experience with business process analysis and modeling. Commenting on the importance of students becoming familiar with the concept of product life cycles, an information architect and CEO of a communications consultancy had the following to say:

> They should be able to think about a product and ask, What's the business problem? What do the people who actually do the work [that the product supports] say? When the customer orders the product, what happens? What humans have to intervene? What's the output from that process? And so on. Students ideally would be able to diagram a product life cycle. Again, we're getting at process and within process, little procedures. Some of them done by software, some done by humans. But there's a question of scale. At the large scale, there's a customer [who] orders a product. At the lower level, there's a procedure where a customer's credit is checked and so on. It helps, I think, for students to differentiate between [simple processes] and large processes, which have within each process dozens of little procedures.

Understanding content development as a multifaceted process intrinsically connected to a larger ecosystem responsive to business goals and business and product development processes is an important competency for technical communication students to develop as part of their preparation for work in the content discipline. This understanding is often assessed as part of the interview process for entry-level content-focused positions, though most hiring managers do not expect to see strong evidence of this competency among recent college graduate candidates (although they would certainly welcome it).

Evidence of systems knowledge and thinking in job applications and interviews should attract the attention of hiring managers for entry-level content-focused positions. As educators who want graduates of our programs to gain meaningful employment as technical communicators, we should pay attention to new or evolving required and desired competencies for entry-level positions in the profession. While systems knowledge in different contexts looks different and is largely gained through on-the-job experience, we take the position that faculty should be preparing students for career mobility, for securing entry-level positions and then moving *quickly* into more advanced, strategic positions that afford them agency that leads to impact.

When we asked industry leaders how entry-level employees could move into more strategic roles and about what skills and competencies people in those roles need, they all spoke about the need to think business-wide, cross-team, and in terms of systems and abstractions, structure, and semantics. Teaching students how to build a business case for process improvements was the biggest theme in response to our question about how educators can best prepare students to be leaders and innovators. The ability to do this requires knowledge of product and content life cycles. It requires a system-level understanding of how products are developed and how customer product journeys influence development decisions and decisions around the life cycle of and possibilities for content connected with a product. A senior information architect who also taught part time in a technical writing certificate program talked about the importance of teaching students to think in terms of business strategy and customer product journeys. They need to become practiced, he said, in asking questions such as "Why does someone buy your products? Why do they choose your solution over someone else's and how can you help them choose the right solution? How can you support that decision?" Students need to think beyond writing content for performing particular tasks to the bigger picture—the whole context in which users are working.

Importantly, industry leaders encouraged educators to help students see and understand how to connect the dots between their work and the success of the business, including the success of digital content transformations. One industry leader, president of a content professionals employment agency, had the following to say about preparing students to be future leaders and innovators:

> It's communication skills. It's how to build a business case. It's, how could we be doing this better? Now, we're getting into process improvement. Are you teaching process improvement and business process modeling as part of your technical communication program? I doubt it. Yet, these are the very skills [students need] coming out of academia to be [future] thought leaders.

Another leader, president and owner of a content solutions consultancy, also stressed the importance of teaching students to build business cases:

> To identify processes that can be improved, one needs [to] gather data that demonstrates inefficiencies and other negative impacts. One then needs to be able to demonstrate through data the impact of proposed process improvements. Building a business case means using data to demonstrate need and impact.

This industry leader also talked about how students need training in how to recognize process improvement opportunities and identify use cases that can help them build a business case:

> Look, this company, by cleaning up their SEO [search engine optimization], was able to reduce support calls by this much," or, "This company,

by moving to DITA, was able to achieve this amount of reuse," and blah blah blah." If you bring an executive or director a new idea, they will want to know, "Well, has it been proven? What kinds of expectations can we have?" And so having some of that already at their fingertips through some of their education ... I think it is critical to become a leader, to be able to find that data...to know what impact these things might have, and who they appeal to and those types of things.

In Chapter 7, we articulate ways in which educators can cultivate systems thinking in students and build curricular pathways that expose students to business processes and management practices as well as product development processes and product life cycles.

Technical Knowledge

In responses to questions about hiring needs for entry-level content-focused positions, as we articulated in Chapter 4, industry leaders mainly emphasized the need for content professionals who have technical knowledge and skills. Key themes that came up over and over in discussions of hiring needs were as follows:

- Ability to grasp technology fast
- Eagerness to use technology to problem solve
- Proficiency with computer science concepts and programming
- Expertise in a technical domain.

We were not surprised, then, when technical knowledge and skills were emphasized again in response to our question about the most important skills and competencies that students should learn in undergraduate programs aimed at preparing content professionals. Both industry and academic leaders agreed that this competency was not about learning particular tools but about learning how different tools work, how to work in different kinds of application environments, and how to move between and among tools to achieve different goals.

Tools that support and facilitate structured authoring workflows, including markup languages, were stressed as the most important for students to gain familiarity and experience with. These include, among others, markup languages such as XML and HTML; data interchange formats such as JSON; lightweight markup languages such as Markdown and AsciiDoc; editing platforms or build systems; component content management systems; version control systems such as Git or GitHub; and publishing processors, particularly those built on the DITA Open Toolkit. Students also need familiarity and experience with project management and Agile project management tools. (Note that, in Chapter 7, we do not advocate for prioritizing the teaching of code in curricula; rather, we advocate for creating opportunities for students to navigate messy situations and move between and among technologies and tools to solve problems.)

Industry leaders emphasized the importance of students not only learning how to work in different kinds of environments and how to move between and among tools but also developing hardware and software knowledge. This is because an understanding of hardware and software—what they are, how they are built, how they are intertwined—is necessary for documenting complex products from beginning to end. Writers need to understand the life cycle of a product and how the content they write fits into it. The founder and CEO of a content solutions consultancy offered the following explanation of why software and hardware knowledge is increasingly important for content professionals:

> Products are becoming smart, and that means that ... products now have actually software in them and electronics. So you can no longer rely on having to deal with machinery only or with software only. It's all interweaved, and it's both hardware and software in a product. That means you must be able to work with both to understand both.

In addition to products becoming smart, products are becoming modular and variant rich. Customers can now use a product configuration engine to configure a product to meet their needs. This means that content, too, must be modular and variant rich and must account for all possible product configurations. Content professionals thus must intimately understand the hardware and software components of a product to create technical documentation.

The best place for students to gain this knowledge, suggested a senior director of user assistance for a major software solutions company, is through engineering and computer science and programming classes. She explained that students do not need to take these classes because they are going to be programmers, but because they need to understand "that end-to-end process of how software, how a program, is eventually in the hands of a consumer." She continued,

> Think of any application that you use, even a browser—I'm on a computer, but how does that browser get to me? How does somebody decide how it's... What was the concept? What was this idea? How do you go from idea to delivery to consumer, and how does the consumer consume it. That process, from conception to delivery, to even obsolescence,

I think is something that everyone should understand. That's something that's definitely missing sometimes in entry level writers.

Domain knowledge, after knowledge of tools that support and facilitate structured authoring workflows and after hardware and software knowledge, was the third dominant theme among responses to our question regarding the most important skills and competencies for students interested in technical content-focused careers to learn. Content professionals with technical qualifications and specialized knowledge, particularly in a STEM field, are highly desirable because of their ability to understand complex technical information quickly

Communication Skills

It makes sense to acknowledge that students completing degrees in technical communication need to develop strong communication skills. For students preparing for careers as content professionals, these skills are as important without a doubt. Not only do content professionals need to craft arguments (e.g., build business cases, advocate for process improvements), tailor communications to specific audiences for specific purposes, and navigate often messy and complex social situations, they need to become proficient in analyzing and responding to vastly expanded rhetorical situations and in engaging in conversations around content at a cross-team and system level.

As we discussed in Chapter 2, for technical content to be usable, useful, and accessible at scale, in all the ways both users and stakeholders want, it must be modeled and engineered before it exists. Content professionals at all levels are increasingly contributing to decisions around how to model and engineer content, including writers in Agile workflows tasked with supporting a product or series of products. For example, writers must understand user journeys and use cases and user needs and questions, and they need to be able to align this knowledge with content models and develop content accordingly. The activity of audience analysis, then, extends beyond the lone writer analyzing intended users of a document and then making decisions about what to write, how to write it, and how to format it. A key outcome of audience analysis in the context of structured authoring workflows is identifying what users need to know and what questions they have, and then mapping those needs and questions to modular content units for development—units that conform to content models and that are typed, designed for reuse, and structured in accordance with a schema or template.

In describing the most important skills for students to develop, many industry leaders focused on the activities of audience analysis in the context of structured authoring workflows. Comments below are representative of the kinds of skills in audience analysis that these leaders said were important for students to develop:

> What exactly is it that we're doing? And that leads to things like, okay, user and task analysis. Who are these people we're talking to, what do they need, what questions are they asking because each question or each type of question comes with a requirement for a particular block of content as defined formally in the schema or the DTD. So, tech comm is this process of mapping the universe of what your audience needs to know [and then] breaking it down as appropriate into [typed topics].
>
> —*CEO and founder of content solutions consultancy (interviewee)*

> The difference [between traditional and modern content development processes] really is that now you have to start with the user.... You have to know who your users are and what they need and when they need it. Again, it always goes back to the right information at the right time, and the right amount of information in the right place.
>
> —*Senior director of user assistance for software solutions company (interviewee)*

> But it's that concept of understanding your audience to give them just enough information to answer their question. That should be embedded into the brain of every single technical communicator coming out of their first class or first semester.
>
> —*President of content professionals employment agency (interviewee)*

Structured content is modular, typed content that responds to specific user needs and questions—and offers only the information users need in a specific moment of use. To determine what content to write and how to appropriately type, chunk, and write it for reuse, content professionals must engage in audience analysis and, in many work contexts, align content development decisions with existing content models, often developed by content strategists. For entry-level content professionals working in organizations that have adopted digital content transformation workflows and Agile practices, developing successful technical content requires being tuned into the larger system of content operations and product management and development.

The ways in which writers, or content professionals, need to think about audience analysis changes in structured authoring versus more traditional publishing environments. An academic leader who has published widely on aspects of the content discipline talked about how today's audiences "are much more complicated than audiences as they're talked about as a rhetorical concept":

> We can't just assume that we're talking to a singular audience. We're talking about an audience that is a part of the community and that the community is part of that audience as well. It's not an audience that shares the same culture as you do. It's not an audience that shares the same technical knowledge as you do. So how do you begin to take that audience knowledge and say, let's make it just a bit more complicated by expanding it, deepening it, multiplying it, and doing something with it?

To understand audiences in the way that this academic leader suggests, content professionals need to be skilled in conducting research and gathering and analyzing data. This ability was emphasized by a number of leaders, as data can go a long way in helping content teams understand users and their journeys with a product and product content. Being well versed in the full data life cycle of collection, analysis, interpretation, and application was a skill set that several leaders identified as important for students to develop.

Strong interviewing skills were particularly stressed by both industry and academic leaders—students need to be practiced in talking with people and knowing how to find answers to important questions and make sense of those answers for particular purposes:

> Certainly how to do research, how to find the right information, how to interview subject matter experts, how to think critically—those are the major skills.... I think those skills are underserved these days.
>
> —*Information architect, former STC fellow, and technical communication certificate program instructor (interviewee)*

> We know from everything we see or everyone we talk to [that conducting research] is such a huge part of what a professional does—that ability to interview people in particular, but also to know how to do a small scale survey and what to do with the results when you get them. So how can you analyze those results in a way that demonstrates the value that they add to whatever the problem is that's being addressed and how do you communicate that to an audience visually, probably, so people understand what the data means and why they should pay attention to it?
>
> —*Professor and director of corporate relations for a department of technical communication (interviewee)*

> I'd go back to good interviewing skills and things like that—to be able to dissect it down enough so that they understand it and sound like

they're knowledgeable when they write their content. But at the same time, recognize that what they had to go through to learn the content is not necessarily what they need to put into the content, that they are not their audience, and so sometimes they have to ask a whole bunch of questions to understand something and then dissect that down into something for a different audience.

—*President and owner of a content solutions consultancy (interviewee)*

Audience analysis, research (collecting and analyzing data, conducting interviews, designing and administering surveys), and oral communication (both delivering presentations and articulating oral arguments in meetings and one-on-one interactions) continue to be important skills for technical communication undergraduate programs to cultivate in students; for programs developing curricular pathways that prepare content professionals, these skills require the ability to analyze and respond to vastly expanded rhetorical situations and to engage in conversations around content at a cross-team and system level. Communication skills such as these are also associated with career mobility. Both industry and academic leaders, in response to our question about preparing students to be leaders and innovators in the discipline of content, emphasized the importance of communication skills, in addition to systems knowledge and thinking and technical knowledge, for moving from entry-level to more advanced content positions.

Collaboration Mindset

While not identified as most important as often as structured authoring, systems knowledge and thinking, technical knowledge, and communication skills, a collaboration mindset was a theme that cut across responses to our question about skills and competencies students should learn in undergraduate programs that aim to prepare content professionals.

This mindset is critical to the success of today's content professionals, from entry to advanced levels, given the intensely collaborative and distributive environments in which they now work. Two prominent content solutions consultants who had overseen multiple digital content transformation initiatives in organizations described the importance of a collaborative mindset in these ways:

And you know what, they're going to have to depend on other people. And sometimes those people are going to let them down, and they need to know how to react to that. They need to know how to be professional

> in that circumstance. They need to know how to lead. They need to know how to continue to deliver or go back and negotiate.
>
> —*CEO and founder of a content teams and leaders training consultancy, former STC fellow, and program chair of a technical writing and communication certificate program (interviewee)*

> Team building skills, right? The next generation technical communicator needs to be a team builder, right? Needs to be able to really exercise those social skills to build teams within their writing teams, within their product teams, within their user communities, within their industry. They need to develop better social skills and be able to step out of their skin. And need to have strategies to be able to foster that. I think that would be the biggest thing I would suggest that students coming into the industry for the first time need to be aware of. They need to understand that being active in the community is an important part of being part of the profession.
>
> —*Principal consultant and co-founder of an authoring solutions company (interviewee)*

Cultivating a collaborative mindset in students and creating opportunities for cross-disciplinary interactions through applied projects is also a way programs can prepare future leaders and innovators in the content discipline, as stressed by these two industry leaders:

> And that's a strategy as well, of really from a leadership perspective, is how do we collaborate with these different organizations, again, what value do we bring to those other organizations, how can we help them, how can they help us, and understand that, really, we all have the same goal in helping the user through their customer journey, whether that's being exposed to the product in the first place, or it's learning to use it, or whatever. And so that collaboration piece… I think it's critical for a leader to know how to collaborate and work with the departments that are not their own.
>
> —*President and owner of a content solutions consultancy (interviewee)*

> But I don't see many programs that are looking at things like cross functional leadership, for example. How do I work on a team with engineers and this and that, and actually drive a conversation, get my opinion out in an appropriate way and have data to back it up and actually express it in a way that people accept it and we move forward with the direction I think is important for the user, etc.? So I think that

is the kind of set of skills [that we need]. It's softer skills that I think are missing.

—*CEO and founder of a content teams and leaders training consultancy, former STC fellow, and program chair of a technical writing and communication certificate program (interviewee)*

A Competencies Framework for Technical Content Professionals

How educational programs can cultivate critical pragmatic practitioners who contribute in impactful ways to the content discipline is a key question at the heart of this book.

Critical pragmatic practitioners in entry-level content-focused positions in the content discipline are systems thinkers who can solve problems in innovative, proactive manners; who want to learn and experiment with new approaches and possibilities; and who seek out opportunities to share knowledge and ideas and to gain consensus with stakeholders on best pathways forward. These practitioners are what Reich (1992) described as symbolic-analytic workers—their skills in systems thinking, abstraction, experimentation, and collaboration are essential to performing well as content professionals in organizations that have adopted digital transformation workflows.

As educators, we are interested in how our educational programs can prepare students for mobility in the content discipline—to first-land entry-level content-focused positions and then, through experience and growth, move into more strategic roles that afford them opportunities to meaningfully contribute to the digital content transformation initiatives of their organizations. As we argued in Chapter 1, content professionals who are prepared to assume strategic roles are needed more than ever. Whereas entry-level content professionals can and do advocate for users and stakeholders within the limited context of the product areas that they oversee and the content strategy frameworks that guide technical content development processes, content professionals who assume strategic roles are elevated from user advocates to champions—a point we emphasized in Chapter 3. As champions, content professionals can assert agency in the ways in which products and users interact by means of engaging with the processes by which products are designed and developed. So too are they well-positioned to define and drive initiatives and activities with social justice goals, gaining buy-in from management and other leaders and assessing impact through data-driven methods.

This is an opportune moment for technical communication educators to develop curricular pathways that prepare students as viable candidates for entry-level content-focused positions that, as we explained in Chapter 4, are often offered to candidates in STEM fields, particularly computer science. When our graduates are not competitive for entry-level content-focused positions, they do not get in the door. They thus cannot impact in meaningful ways organizational initiatives to transform business processes and customers' experiences through the adoption of new digital technologies, including technologies that afford powerful ways of creating, managing, and delivering content at scale.

While we acknowledge that it's not realistic for new college graduates to be hired into strategic positions right out of college, we also acknowledge that we can prepare students as critical pragmatic content professionals with a mobility skillset—one that they can build upon as they gain requisite on-the-ground experience. Rahel Bailie and Noz Urbina (2013) suggested that technical communicators whose roles have been primarily as content authors are not equipped to take on more strategic roles like content strategists. Without additional training and experience, they argued, it is unrealistic to expect authors

> to know enough about content standards and content modeling, re-use models, metadata best practices, microformats, writing for syndication, and componentization for content management systems to make informed decisions about how to pull all of the pieces together to support business directives.
>
> *(p. 254)*

A content strategist, in their view, is a senior position requiring years of on-the-ground experience with various aspects of planning and implementing a content strategy. Similarly, an information architect, as Ann Rockley (2010) and Joe Gollner (2010) have argued, often requires years of on-the-ground experience and a comprehensive understanding of an organization's content in various forms. We argue, based on what we have learned through our conversations with industry leaders and our experiences as technical communication educators, that students can be prepared as critical pragmatic content professionals with a mobility skill set that equips them to adapt, learn, and innovate in modern content ecosystems and move into more strategic roles after gaining requisite organizational knowledge and experience.

What does this mobility skillset look like? Drawing on our research findings, as well as those of technical communication scholars who have identified and described needed competencies and skills through literacy frameworks, analyses of job advertisements, interviews, surveys, systematic reviews, and experience-informed perspectives (refer to Chapter 5), we introduce a competencies framework for content professionals (Figure 6.1) that expands Henschel and Melonçon's explanatory matrix of conceptual and practical skills for technical communication undergraduate curricula.

Henschel and Melonçon's matrix is helpful for thinking through the skills and competencies needed to understand, contribute to, and innovate within complex information systems, including those connected to digital content transformations. We particularly appreciate how their matrix as a competencies framework attends to the need to cultivate graduates who can readily move from entry-level positions into more strategic positions. In their matrix, Henschel and Melonçon identified five conceptual skills—critical system thinking, rhetorical proficiency, abstraction, experimentation, and social proficiency—that represent "the higher-order knowledge and literacies a technical communicator needs to be successful and remain flexible in the ever-changing workplace" (p. 5). They placed critical system thinking at the center of the matrix as a way to show how it anchors the other four skills and how students need to become "critical system thinkers who can deploy any one or a combination of the conceptual skills as needed to solve problems in the workplace" (p. 9). They also identified practical skills that correspond with conceptual skill areas.

In Figure 6.1, we identify the conceptual skills that entry-level content professionals need to perform well in the content discipline (and increasingly need to gain access to entry-level positions). These skills are largely the same as those identified by Henschel and Melonçon, but what they look like in the

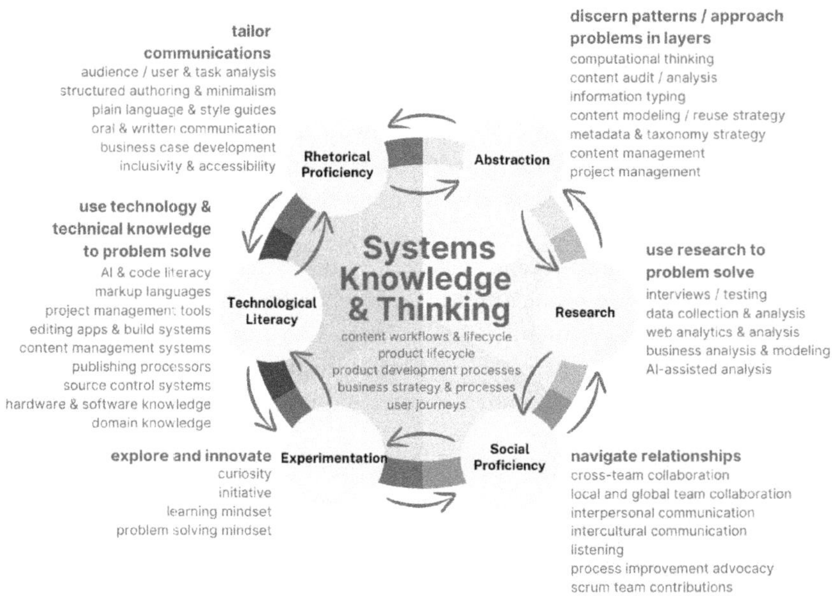

FIGURE 6.1 Competencies Framework for Content Professionals.

context of digital content transformation workflows is different from what they look like in more traditional document-based publication environments. In our competencies framework for content professionals, we do not list practical skills that correspond to specific conceptual skills; rather, for each conceptual skill, we list associated activities, approaches, technologies, and/or mindsets with which content professionals must engage. As was the intention of Henschel and Melonçon in listing practical skills, listing these associations offers a way for educators to make sense of the parts of the whole of what it means to be a college-educated content professional today.

This framework offers technical communication educators and researchers a way to talk about the conceptual skills and associated activities, approaches, technologies, and mindsets needed to perform knowledge work in the content discipline as an entry-level and mid-level content professional. The conceptual skills also map well to those most mentioned among the six studies presented in Chapter 5 (refer to Table 5.2), those articulated in competency discussions connected to the content discipline, and those articulated in multiple literacy frameworks for technical communication undergraduate curricula. In various instances, technical writing and information design are highlighted as important skills; in modern content ecosystems, these skills fall under the umbrella of structured authoring.

Figure 6.1 identifies seven conceptual skills for content professionals. **Systems knowledge and thinking** are critical for understanding the rhetorical context of content ecosystems. Performing well in this context requires **rhetorical proficiency**; skills in **abstraction** and **research**; **social proficiency**; an **experimentation** mindset; and **technological literacy**. We discuss these skills in subsequent sections. But we also want to call attention to another way of conceiving the framework (a suggestion offered by an industry peer and friend). In thinking in terms of a Venn diagram, one might put rhetorical proficiency, technological literacy, and social proficiency as the three primary competencies represented in the diagram, with abstraction, research, and experimentation viewed as key activities bridging or connecting these competencies (abstraction between rhetorical and technical, research between rhetorical and social, and experimentation between technical and social). At the center of the diagram, anchoring the primary competencies would be the emergent knowledge area of systems knowledge and thinking. This conception of the framework allows for thinking of the seven conceptual skills as three primary competencies, three key activities, and one emergent knowledge area.

Systems Knowledge and Thinking

Systems thinking involves the ability to understand how different parts of the whole are related. When presented with a problem or an issue, as

Reich (1992) described, systems thinkers can "discern larger causes, consequences, and relationships" (p. 30); they can redefine problems in terms of a "broad system of forces, variables, and outcomes" that might not be immediately apparent (p. 31). In Henschel and Melonçon's matrix, critical pragmatic practitioners are those who "understand the processes by which parts are linked together" and can "consider ideological/power stances of those structures and critique when necessary" (p. 10).

Systems knowledge and thinking in the context of the content discipline is about having a big-picture view—or systems view—of the processes, workflows, and life cycles that shape content activities and roles and the nature of content itself. It is about understanding interconnected systems and their parts, what we might call the new rhetorical situation that itself encompasses multiple rhetorical situations. Content professionals must be prepared to work and think at a system level and to engage stakeholders from across an organization; so too must they be prepared to apply rhetorical and technical expertise as well as social proficiency to both understand complex products and systems and contribute meaningfully to content ecosystems.

Students who want to work as content professionals and who want opportunities to advance in their roles and move into other disciplinary areas of the content discipline need to develop systems knowledge and become practiced in systems thinking. In the curriculum of a program in technical communication and related fields, this means learning about the content discipline, with a particular focus on content strategy and content operations, to understand interconnecting systems supporting content workflows and the life cycle of content. An academic path supporting this exploration perhaps should include or encourage courses taken as electives or connected to a minor or double major to learn about business management and processes and product development processes and life cycles. Systems thinking and knowledge also inform understandings of user journeys with a product and content connected to that product. Content marketing and user experience design as adjacent fields can offer students rich insight into user journeys and strategies for researching and modeling journeys.

Rhetorical Proficiency

Rhetorical proficiency situates the ability to "compose content for a variety of audiences and purposes" (Henschel & Melonçon, 2014, p. 10) in the context of modern content ecosystems. Structured authoring is thus an essential practical skill associated with rhetorical proficiency. Carlos (Evia, 2018), in fact, advocated for rhetorical effectiveness as a sixth characteristic of intelligent content (adding to the oft-cited five characteristics articulated by Rockley, Cooper, & Abel, 2015). He argued,

> Intelligence applied to content development and delivery does not guarantee intelligence in reconciling the needs and perspectives of readers and writers. Rhetorically effective content can connect topical and authorial intentions with user responses regardless of a deliverable's format of context. If content production can be automated, it still depends on humans as authors, information architects, and experience evaluators in order to be rhetorically effective.
>
> *(p. 157)*

Students being prepared as content professionals should be practiced in creating intelligent content (what we have referred to as technical content throughout the book), and they should have a conceptual understanding of structured authoring as a methodology as well as be familiar with the standards, schemas, and technologies available for supporting structured authoring. Structured authoring as a rhetorical, computational, technological, and social process is first and foremost driven by user experience advocacy goals, which cannot be divorced from rhetorical proficiency. Students thus need to develop audience and task analysis skills in the context of structured authoring workflows. They need to be able to identify through research what users need to know and what questions they have and then map those needs and questions to modular content units for development—units that conform to content models and that are typed, designed for reuse, and structured in accordance with a schema or template.

As a rhetorical, computational, technological, and social process, structured authoring requires the ability to employ a layers of abstraction approach to authoring; to apply knowledge of markup languages; to learn and work with the authoring, management, and processing tools of different content stakeholder groups; and to collaborate with product, development, UX, and content teams on product and content design decisions.

Given the ways in which composing processes are interconnected with larger systems, processes, and stakeholders (e.g., product managers, developers, director of content operations), the ability to effectively communicate with stakeholders and advocate for process improvements and teams is essential to performing well in modern content ecosystems. Rhetorical proficiency as a conceptual skill additionally focuses on the ability to craft arguments (e.g., build business cases, advocate for process improvements), tailor communications to specific audiences for specific purposes, and navigate often messy and complex social situations. Content professionals proficient in this skill can analyze and respond to the vastly expanded rhetorical situation of modern content ecosystems and engage in conversations around content at a cross-team and system level.

Abstraction

Henschel and Melonçon described the conceptual skill of abstraction as the ability to "discover patterns and meaning, rearrange information in new ways" (p. 10); likewise, Johndan Johnson-Eilola (1996) described it as the ability to "discern patterns, relationships, and hierarchies in large masses of information" (p. 260). In modern content ecosystems, this ability is required for creating and managing technical content at all stages of its life cycle, from performing a content audit to typing and modeling content based on user research and business needs to managing content shared across teams and products.

Abstraction is linked to systems knowledge and thinking. Carlos (Evia, 2018) described the concept of abstraction in terms of *computational thinking*. Abstraction is about approaching problems in layers, working on each layer individually, and then recombining the layers to solve the problem. Carlos encouraged a view of the content deliverable not as a static object but rather as a dynamic process and a product of computational thinking in layers of abstraction (p. 149). Content authors, like computer engineers, he suggested, need to approach intelligent content in layers of abstraction. This computational thinking practice encourages authors to "view content as non-linear and envision more than one outcome based on principles of reuse and single-sourcing" (p. 151). The conceptual skill of abstraction, then, is a way of thinking about content—about its potential future applications and final outcomes (p. 152).

Abstraction as a skill is essential for carrying out content-focused activities in organizations that have adopted digital content transformations. This skill requires systems knowledge and thinking and is interconnected with rhetorical proficiency and the ability to use research to problem solve.

Research

Henschel and Melonçon identified research as a practical skill connected with the conceptual skill of abstraction in their explanatory matrix. Because solving problems in content ecosystems requires research, and because content professionals need to collect various kinds of data depending on the problem that they are trying to solve, we elevated research to a conceptual skill in our competencies for the content professional framework. Research activities that content professionals frequently carry out include conducting interviews and usability tests, collecting data that helps content teams understand users and their journeys with a product and product content, and analyzing and modeling business processes to identify opportunities to improve processes and workflows, including content workflows.

The ability to use research to solve problems requires skills in abstraction as well. Collecting and analyzing data for the purpose of problem-solving means discerning patterns and discovering meaning. Using research to solve problems also requires systems knowledge and thinking in the context of content ecosystems, and it requires social proficiency—the ability to navigate relationships toward the end of collecting useful data and analyzing and modeling business processes. Eva Brumberger and Claire Lauer (2015), in their analysis of close to 1,000 US job postings, identified research as one of the top five professional competencies that technical communication employers seek in job candidates—as such a critical skill for performing well in content ecosystems, we position research as a conceptual skill in the competencies framework for content professionals.

Social Proficiency

Social proficiency as a conceptual skill focuses on the ability to "collaborate, negotiate, and achieve consensus" (Henschel & Melonçon, 2014, p. 10). In Agile development environments where a majority of content professionals now work, daily collaboration and communication with members of cross-disciplinary teams are the norm. Content professionals can expect to give daily oral presentations on scrum teams and frequent presentations to project stakeholders, from status updates to process proposals to product or content design needs. Collaboration with members of cross-organizational departments, such as marketing, training, and engineering, is also expected.

Social proficiency is about navigating relationships within and across local and global teams. Content professionals need to be ready to share their knowledge, ideas, needs, problems, and solutions with stakeholders; to be able to accept criticism and solicit help; and "to discern what others need and view things from others' perspectives" (Reich, 1992, p. 32). This ability requires interpersonal, intercultural, and global communication skills as well as listening skills—that is, knowing when to listen, how to listen, and how to respond appropriately and productively to others' ideas, contributions, and criticisms. We particularly emphasize the work of researchers like Godwin Agboka and Isidore Isidore Dorpenyo (e.g., Agboka & Dorpenyo, 2022) for framing discussions related to social justice and global communication as pillars of social proficiency.

Whereas Henschel and Melonçon located communication skills within social proficiency, we located oral and written communication skills within rhetorical proficiency and interpersonal and intercultural communication skills within social proficiency—these conceptual skill areas are intrinsically linked and cannot be divorced from systems knowledge and thinking or taught in isolation from skills in abstraction and research. As one can begin to see, and as Kelli Cargile Cook (2002) argued, conceptual skills—or what they called "layered literacies"—are not isolated and must be taught as integrated (p. 24).

The conceptual skill of social proficiency is as important as ever for content professionals who must continually navigate relationships to achieve goals for content. Not one industry leader or alumni with whom we spoke worked in isolation, nor did any of the content professionals they referenced in interviews or survey responses.

Experimentation Mindset

An experimentation mindset is about the willingness to try new approaches and technologies for the sake of learning, innovating, and identifying ways to improve processes. It's about having the mindset that lifelong learning, adaptability, and curiosity are central to what it means to be an impactful professional in the content discipline.

Succeeding as a content professional in organizations that have adopted structured authoring workflows requires taking on challenges and learning from them. Continuous technological changes and continuous product development cycles mean content professionals must be curious and adaptable and be willing to tinker with technology, experiment with new ways of doing things, take initiative, and ask for guidance when needed. When presented with messy tasks, content professionals need to be able to quickly get to the work of problem-solving and leveraging available resources toward a solution.

An experimentation mindset is a quality that educational programs can cultivate in students through problem-solving projects, exploration of different technologies and tools, experiential learning, and interdisciplinary collaborations.

Technological Literacy

Henschel and Melonçon (2014) identified "technological literacy" as a practical skill located in the conceptual skill category of experimentation. Given the different levels of technological literacy described by Hovde and Renguette (2017) and the growing need for students to develop both code and AI literacy, and given the technical knowledge and skills identified as important by our research participants, we elevate technological literacy to a critical conceptual skill for content professionals.

Importantly, technological literacy is not about knowing particular tools but about knowing how to overcome messy situations and move between and among technologies and tools to solve problems. It is also about knowing how different content technologies work, how hardware and software work, and how programming and markup languages work. We recognize Marjorie Rush Hovde and Corinne Renguette's (2017) four levels of technological literacy as particularly relevant for content professionals—functional, conceptual, evaluative, and critical. So too is code literacy, and we find the three facets of code literacy identified by Ann Duin and Jason Tham (2018) helpful in making sense of this competency: code-as-language, code-as-tool, and code-as-structure. These facets can be mapped onto the four levels of technological literacy. Peter Cardon et al. (2023) argue that communicators (including educators, students, and practitioners) should develop AI literacy, which in their model involves four sets of capabilities: application, authenticity, accountability, and agency. They add that, for students, "Business communication courses are among the most important courses to develop it given the applied focus on communicating in the workplace" (2023, p. 277).

Students need to gain experience using tools, such as editing applications and version control systems, and authoring in languages, such as XML and Markdown, that support technical content. As important, and arguably more important, however, they need to develop conceptual knowledge of how AI, programming languages, markup languages, and content tools and systems work. Students additionally need to develop conceptual knowledge of how hardware and software work, particularly as smart technologies evolve. An understanding of hardware and software—what they are, how they are built, how they are intertwined—is necessary for documenting complex products from beginning to end.

Instead of becoming hyper-specialized in one specific tool or software application, the goal should be to develop the ability to transfer knowledge of one language, tool, or system to another—to move between and among different application environments.

Curricular Gaps and Opportunities

Our goal throughout this project has been to better understand changing activities, roles, and hiring needs on technical content teams and their implications for technical communication education. We have been particularly interested in identifying competencies that afford students mobility—that afford them opportunities to lead and innovate and to advance into more strategic positions within their organizations and the profession.

These are the competencies that we can associate with agency or the kinds of critical discourse workers for which Greg Wilson and Rachel Woldford

(2017) urged the field to cultivate. They argued for moving our educational programs away from a focus on critique, which tends to cultivate critical discursive consumers, to a focus on action, which has the goal of equipping students with skill sets that provide them access to workplace discourse—these skill sets should arm them with agency and position them to impact workplace discourse. Wilson and Woldford (2017), drawing on Jeffrey Nealon's (2012) proposed post-postmodern theory, refer to this positioning as "critically informed situated action" and this conception of a post-postmodern discourse worker as an "actualized technical communicator" (p. 14).

Melonçon and Schreiber's (2022) "critical pragmatic practitioner" shares the same idea—that the practitioners we are cultivating in our undergraduate programs should be able to reflect and critique and also "make, produce, and communicate" as a result of their education. We claim that making, producing, and communicating in organizations that have adopted digital content transformation workflows is very different from what these applied activities look like in more traditional document-based information development environments. As a result, the skill sets that aspiring content professionals need to develop are different from those that students aspiring to work in other facets of technical communication may need to develop.

As we highlighted in Chapter 1, few undergraduate programs in the technical communication and related fields in the United States have created curricular pathways that cultivate the competencies (and their associated activities, approaches, technologies, and/or mindsets) identified in Figure 6.1. This gap in technical communication curricula has been called out by numerous scholars in the field (e.g., Batova & Andersen, 2017; Bridgeford, 2020a; McDaniel & Steward, 2011; Gu & Pullman, 2008; Spilka, 2009; Tham, Howard, & Verhulsdonck, 2022). More recently, Jason Tham, Tharon Howard, and Gustav Verhulsdonck (2022) specifically addressed the need for programs to "align closely with industry trends and practices so to inform their curricular design and development" (p. 24), and they articulated strategies and offered educators resources for developing curricula that prepared students to engage with content strategy, AI, and design thinking as mainstream practices (all connected to the discipline of content).

Research on program learning outcomes (PLOs) confirms the need for developing curricular pathways as well. Geoffrey Clegg et al. (2021) examined the PLOs of 47 undergraduate degree programs in technical and professional communication (TPC). Their findings indicated that PLOs across institutions continue to focus on preparing students for document-based authoring environments. The most common outcomes included rhetoric, writing, technology, and design. The authors noted that the outcome "rhetoric" was most often connected to "the primary rhetorical concepts of addressing the final deliverable to a specific audience for a defined purpose"

(p. 25). The outcome "writing" was most often connected to documents, though some institutions, as the authors note, "offer a more complete interpretation of the kind of writing that TPC programs are preparing students to do in the workplace," including writing effectively in multiple forms such as print and mixed media. Whereas "technology" as an outcome focused on both the ability to use and critique different tools and technologies, "design" as an outcome focused largely on application of visual design principles and visual communication strategies to documents and media objects (p. 26). Programs had certainly expanded outcomes to include a focus on multimodal and multimedia composition, but they had yet to embrace outcomes that prepare students for working in organizations that had adopted or were planning to adopt structured authoring workflows.

Calls for curricular innovation in the field are not just coming from academic leaders. Industry leaders, too, have been asking how future generations of content professionals will be prepared to contribute to and help shape digital content transformation initiatives. For example, in an episode of the podcast *Towards a Smarter World* focused on the benefits of content models, host Cruce Saunders (2022) talked with Kate Kenyon, Head of Content at Chase International Consumer Bank, about the critical role of content people in defining and overseeing content design and content engineering activities. Saunders commented:

> We have been in the industry long enough that we know there needs to be a next generation and that process is not necessarily evident in terms of the educational system. How are content professionals of the future born and made now? And how should they be?

Kenyon responded with an anecdote of a conversation she had just had with a friend who was an expert in content modeling. That friend was concerned about the dearth of available people prepared to engage in content modeling and had asked Kenyon, "Where are the people I'm supposed to be teaching this to? I can't keep doing this forever. I need someone to hand this to." Saunders suggested that the industry needed to develop its own professional development approach given how the educational system had not caught up.

Responding to the need for curricular evolution in the content-focused facet of our field's collective identity, scholars have published books, chapters, and articles that offer guidance in teaching structured authoring, content strategy, and other foundational activities at the course or module level (see, e.g., Bridgeford, 2020b; Evia, 2018; Getto & Flanagan, 2024; Getto, Labriola, & Ruszkiewicz, 2022; Robidoux, 2008; Tham, Howard, & Verhulsdonck, 2022). For example, contributors to Tracy Bridgeford's (2020a) edited collection on *Teaching Content Management in Technical and Professional Communication* articulated various approaches to teaching content management at the course level. Tham, Howard, and Verhulsdonck (2022), too, articulated strategies for teaching content strategy, artificial intelligence, and design thinking in technical communication courses—three areas of practice of growing importance in the content discipline. While these contributions to the field have offered educators concrete ideas and guidance for developing course content, they have not had the purpose of describing what curricular innovation might look like at the undergraduate major, minor, certificate, and course levels in different institutional contexts.

In Chapter 7, we take up this task, offering administrators and educators guidance on how they might build curricular pathways that prepare students for work in the content discipline and that are sensitive to different program strengths, foci, and constraints. Practitioners and managers, too, will find strategies for partnering with university programs.

References

Agboka, G. Y., & Dorpenyo, I. K. (2022). The role of technical communicators in confronting injustice—everywhere. *IEEE Transactions on Professional Communication*, *65*(1), 5–10.

Bailie, R. A., & Urbina, N. (2013). *Content strategy for decision makers: Connecting the dots between business, brand and benefits*. Laguna Hills, CA: XML Press.

Batova, T., & Andersen, R. (2017). A systematic literature review of changes in roles/skills in component content management environments and implications for education. *Technical Communication Quarterly*, *26*(2), 173–200.

Bridgeford, T. (2020a). Content management: A pedagogical exigency. In T. Bridgeford (Ed.), *Teaching content management in technical and professional communication* (pp. 1–15). New York: Routledge.

Bridgeford, T. (Ed.). (2020b). *Teaching content management in technical and professional communication*. New York: Routledge.

Brumberger, E., & Lauer, C. (2015). The evolution of technical communication: An analysis of industry job postings. *Technical Communication*, *62*(4), 224–243.

Cardon, P., Fleischmann, C., Aritz, J., Logemann, M., & Heidewald, J. (2023). The challenges and opportunities of AI-assisted writing: Developing AI literacy for the AI age. *Business and Professional Communication Quarterly*, *86*(3), 257–295.

Cargile Cook, K. (2002). Layered literacies: A theoretical frame for technical communication pedagogy. *Technical Communication Quarterly*, *11*(1), 5–29.

Clegg, G., Lauer, J., Phelps, J., & Melonçon, L. (2021). Programmatic outcomes in undergraduate technical and professional communication programs. *Technical Communication Quarterly*, *30*(1), 19–33.

Duin, A. H., & Tham, J. (2018). Cultivating code literacy: Course redesign through advisory board engagement. *Communication Design Quarterly Review*, *6*(3), 44–58.

Evia, C. (2018). *Creating intelligent content with Lightweight DITA*. New York: Routledge.

Getto, G., & Flanagan, S. (Eds.). (2024). The how-to of content strategy: Teaching, training, and application [Special issue]. *Technical Communication*, *71*(3).

Getto, G., Labriola, J., & Ruszkiewicz, S. (2022). *Content strategy: A how-to guide*. New York: Routledge.

Gollner, J. (2010). About content strategy. Retrieved from https://www.gollner.ca/2010/12/about-content-strategy.html.

Gu, B., & Pullman, G. (2008). Mapping out the key parameters of content management. In G. Pullman & B. Gu (Eds.), *Content management: Bridging the gap between theory and practice* (pp. 1–12). Amityville, NY: Baywood.

Henschel, S., & Melonçon, L. (2014). Of horseman and layered literacies: Assessment instruments for aligning technical and professional communication undergraduate curricula with professional expectations. *Programmatic Perspectives*, 6(1), 3–26.

Hovde, M. R., & Renguette, C. C. (2017). Technological literacy: A framework for teaching technical communication software tools. *Technical Communication Quarterly*, *26*(4), 395–411.

Johnson-Eilola, J. (1996). Relocating the value of work: Technical communication in a post-industrial age. *Technical Communication Quarterly*, *5*(3), 245–270.

McDaniel, R., & Steward, S. (2011). Technical communication pedagogy and the broadband divide: Academic and industrial perspectives. In A. P. Lamberti & A. R. Richards (Eds.), *Complex worlds: Digital culture, rhetoric, and professional communication* (pp. 195–212). Amityville, NY: Baywood Publishing.

Melonçon, L., & Schreiber, J. (2022). Introduction: Promoting a sustainable collective identity for technical and professional communication. In J. Schreiber & L. Melonçon (Eds.), *Assembling critical components: A framework for sustaining technical and professional communication* (pp. 3–16). WAC Clearinghouse: University Press of Colorado.

Nealon, J. (2012). *Post-postmodernism: Or, the cultural logic of just-in-time capitalism*. Stanford, CA: Stanford University Press.

Reich, R. (1992). *The work of nations*. New York: Vintage Books.

Robidoux, C. (2008). Rhetorically structured content: Developing a collaborative single-sourcing curriculum. *Technical Communication Quarterly*, *17*(1), 110–135.

Rockley, A. (2010). Managing content: An intelligent content strategy. In B. Boiko & E. M. Hartman (Eds.), *Information management best practices* (pp. 269–277). Utrecht: Hartman Communicatie.

Rockley, A., Cooper, C., & Abel, S. (2015). *Intelligent content: A primer*. Laguna Hills, CA: XML Press.

Saunders, C. (Host). (July 2022). Kate Kenyon: The benefits of content models with Kate Kenyon – episode 27 [Audio podcast episode]. In *Towards a Smarter World*. [A]. Retrieved from https://towardsasmarterworld.podbean.com/e/the-benefits-of-content-models-with-kate-kenyon/.

Spilka, R. (2009). Introduction. In R. Spilka (Ed.), *Digital literacy for technical communication: 21st century theory and practice* (pp. 1–18). New York: Routledge Publishing.

Tham, J., Howard, T., & Verhulsdonck, G. (2022). Extending design thinking, content strategy, and artificial intelligence into technical communication and user experience design programs: Further pedagogical implications. *Journal of Technical Writing and Communication*, *52*(4), 428–459.

Wilson, G., & Wolford, R. (2017). The technical communicator as (post-postmodern) discourse worker. *Journal of Business and Technical Communication*, *31*(1), 3–29.

7

PREPARING STUDENTS FOR ROLES AND ACTIVITIES IN THE DISCIPLINE OF CONTENT

Code for Content? Not So Much

In a 2012 blog post for the Digital Rhetoric Collaborative, the late technical communication scholar Bill Hart-Davidson encouraged his audience of digital rhetoricians to get back to doing their real job and leave the writing of computer code to other people. Hart-Davidson's post, titled "Code? Not so much," was received with mixed and passionate reactions by his academic colleagues in technical communication and rhetoric. At the time, in academic conferences and syllabi for college-level writing courses, it was common to see references to texts like Douglas Rushkoff's (2011) *Program or Be Programmed: Ten Commands for a Digital Age* that were inviting students across disciplines to learn how to write computer code. Hand-writing code—from basic Ruby or Python scripts to creating websites with hundreds of lines of HTML and CSS code without visual editors—was like a badge of honor for some academics in technical communication and related fields.

Some who saw adopting computer code as essential to writing and communication courses disagreed with Hart-Davidson. Others agreed with him. How did his proclamation impact the intersection of technical communication and the discipline of content in academic conversations? Hart-Davidson was an early adopter of structured authoring workflows for technical content in academia, and in his *WRA 420: Advanced Technical Writing* course at Michigan State University, he had students writing with the DITA standard to structure and publish information back in the early 2010s. We had several opportunities over the years to discuss with

DOI: 10.4324/9781003451730-11

Hart-Davidson implications of structured authoring for undergraduate programs, and we would talk about how the idea of writing in "computer code" (from HTML to XML and even Markdown) was not inviting to many of our colleagues in academia who worked under the umbrella of technical communication.

The good news is that nowadays, with established relations with corporate partners who donate licenses for enterprise component content management systems to our college courses and with generative AI tools that can produce draft templates for topics and stylesheets, instructors introducing students to structured authoring workflows do not need to teach actual computer code. After decades of work in the intersection of technical communication and the discipline of content, we can quote Bill Hart-Davidson and say to our colleagues on the academic side of technical communication: Code? Not so much.

This chapter introduces ideas for situating the intersection of technical communication and the discipline of content in college curricula. We reflect on what we reported in Chapter 6 from our research participants that built our proposed competencies framework for content professionals. Our discussion is also based on our decades of experience teaching technical content workflows in our courses, and it is heavily influenced by conversations and informal interactions with colleagues in industry and academia.

Given the limitations that undergraduate courses and programs in technical communication and related fields face at different institutions, we are careful to avoid generalizations and absolute recommendations. For programs interested in building pathways for students to pursue careers in the discipline of content, we suggest a staged approach based on a whole content life cycle. Rahel Bailie and Noz Urbina (2013) explain that a content life cycle "describes the phases that content goes through from planning and creation to various manipulations to publication to post-publication processing to retirement or iteration through another cycle" (p. 265). Each stage of our suggested approach increases in complexity and engagement with the content life cycle. For each stage, we describe topics to cover and why and offer both resources and suggested activities. Later in the chapter, we also share the advice that our research participants had for technical communication educators on how they could keep curricula aligned with the needs of the content-focused facet of the field. The number one recommendation was to emphasize structured authoring throughout curricula. Other key recommendations included prioritizing industry outreach and partnerships, forming interdisciplinary alliances and integrating interdisciplinarity into curricular pathways, and engaging in professional development.

If Not Code, What Then?

We frequently advise colleagues and students who ask us about approaches to teaching technical content workflows in college-level courses. Most of those conversations have focused recently on a complete content life cycle and the many activities and processes that it includes. We do not recommend focusing on the purchase or adoption of one software package or tool, and we try to move code (even HTML and DITA tags) away from the spotlight. Carlos jokingly tells students and colleagues to run, not walk, if an introductory course or presentation about technical content in 2025 and beyond starts with instructions on how to add the absolute path for the DITA-Open Toolkit (DITA-OT) installation directory to the PATH environment variable or on how to launch the command-line interface or terminal—which used to be the main way to access the DITA-OT. Even when we introduce workflows that follow the docs-as-code philosophy, we emphasize the conjunction *as* and focus more on processes like content authoring, editing, and managing in a repository than on writing raw computer code.

Our discussions of the content life cycle in technical communication are based on the model presented by Bailie and Urbina (2013), particularly on the slightly revised version that Bailie uses in her courses and workshops (Figure 7.1).

Bailie and Urbina contrast traditional content and publishing models (a linear activity with a clear beginning and an end) with contemporary content workflows (a cyclical and iterative process). We argue that content

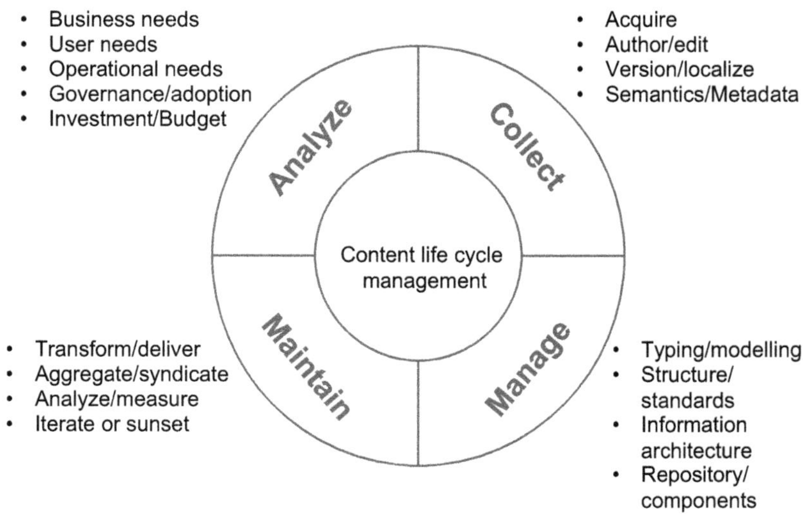

FIGURE 7.1 The Content Life Cycle According to Rahel Bailie.

professionals—and the instructors who educate them in college courses—do not need to understand or be proficient in all the activities involved in the content life cycle. However, they should be aware of the quadrants in the life cycle and their iterative nature.

In our collaborations with Rahel Bailie, we have learned and discussed many considerations about the content life cycle, which include the following:

- Content exists in every quadrant; therefore, any quadrant could be a starting point. Regardless of starting point, content professionals should plan for the activities across the quadrants up front.
- The *Analyze* quadrant includes activities commonly connected with content strategy, which can be applicable to many forms of content beyond technical communication.
- The *Collect* quadrant is where most of the content development, authoring, and creation take place. Sometimes, entry-level professionals are responsible for these activities as content authors. In other cases, advanced professionals are responsible for typing/modeling and structure/standards while entry-level colleagues are responsible for authoring/editing according to those models and defined structures.
- The *Manage* and *Maintain* quadrants involve advanced tasks that rarely would be the responsibility of an entry-level professional, but all content professionals should be aware of the tasks that happen in those quadrants.
- The *Maintain* quadrant can have an audience of human beings (readers or users) or other systems or processes.
- Generative AI, properly implemented and managed, can support and simplify activities across the quadrants.
- The quadrants are subject to variations depending on genres, deliverables, or situations.

Rahel assigns the *Analyze* quadrant to the disciplinary area of content strategy, and then groups the remaining three quadrants in the area of content operations. We propose to have the quadrants of *Collect* and *Manage* also involving the area of content design and the quadrants of *Manage* and *Maintain* related to the area of content engineering. However, we respect the dynamic nature of the life cycle and acknowledge the unavoidable overlap of disciplinary areas and activities.

Table 7.1 presents the disciplinary areas of content that we have mapped for every quadrant of the life cycle.

From a broad perspective, it is easy to see how many (and how few!) of the activities involved in the content life cycle are truly dependent on

TABLE 7.1 Quadrants of the Content Life Cycle Mapped to Main Disciplinary Areas of the Discipline of Content

Quadrant	*Disciplinary Area(s)*
Analyze	Content strategy
Collect	Content operations Content design
Manage	Content operations Content design Content engineering
Maintain	Content operations Content engineering

mastery of computer code. Technical content offers many entry points to interested instructors, researchers, and students in technical communication, but it does require a certain level of technological literacy that is not just about code. The following section introduces ideas for situating the intersection of technical communication and the discipline of content in college curricula. We leave room for the moments and activities that without a doubt require some computer code, but we also build our recommendations on a backbone based on the skills and competencies for content professionals that we introduced in Chapter 6:

- Systems knowledge and thinking
- Rhetorical proficiency
- Abstraction
- Research
- Social proficiency
- Experimentation mindset
- Technological literacy.

There's not much code, and the technological literacy at the center of this section, which we forecasted in Chapter 5, is not about knowing particular software tools or apps. This level of literacy is about knowing how to navigate messy situations and move between and among technologies and tools to solve problems involving diverse groups of human content professionals and the consumers of their information products.

Building Curricular Pathways

We begin this section acknowledging the limitations that undergraduate courses and programs in technical communication and related fields face at different institutions. Technical communication does not have a standardized curriculum across colleges and universities in the United States, and efforts to establish peer accreditation or certification programs have not been well received. Thus, as we propose ideas for connecting the skills and competencies that we reported in Chapter 6 and the content life cycle that we presented earlier in this chapter to curricula, we avoid generalizations and absolute recommendations.

We do not propose complete curriculum reform pathways or critique existing degree requirements. Instead, we assemble recommendations and ideas based on the research that we conducted for this project, our experience teaching about content in technical communication courses for more than two decades, and years of interactions with colleagues in industry and academia who are experts in this domain. Those recommendations and ideas are presented in stages of increasing complexity and engagement with technical communication as it intersects with the discipline of content.

Regardless of course (a single course covering the whole content life cycle or a sequence of courses focused on disciplinary areas of content) or degree (a concentration degree embedded in an English major or a dedicated technical communication major) type, we hope that these broad recommendations and the resources included in this book's companion GitHub repository[1] are useful for instructors interested in exploring the intersections of technical communication and the discipline of content. We also do not want to impose these recommendations as a hyper-pragmatic replacement for a full technical communication curriculum that grounds any approaches in theory. Rebekka has called for curricula with a strong theoretical background that "will help students critically apply these skills to the various technological, cultural, and rhetorical exigencies they will face in (content-related) contexts" (Andersen, 2014, p. 146). She particularly emphasized theoretical curriculum on rhetoric of technology, technology transfer and diffusion, information design, user-centered design, activity theory, actor networks, genre, and human-computer interaction. Carlos (Evia, 2018) proposed coursework on principles of rhetorical problem-solving and computational thinking for technical communication students interested in content-related work. Furthermore, in Chapter 1, we highlighted the need for humanities-trained technical communicators to meaningfully contribute human-centered design values and critical sensibilities to the content life cycle. A participant in the research study behind this book summarized the need for curriculum grounded in the humanities for technical content professionals as follows:

> So, part of it is, one of the implications is that we need to have both the practical kinds of training with specific attention to what is actually going on in industry, with the kinds of questions that's concerning them,

> the kinds of tools that they're using. But also finding a way to talk differently about the humanities roots of tech comm. So that we don't see that there's bifurcation between all of the rhetoric and writing classes that we're taking, or the linguistics classes that we're teaching, and then the workplace. Because I think that's an unproductive division between those two things.
>
> —*Professor and department head (interviewee)*

Although this is not necessarily a how-to book, in its companion GitHub repository, we will maintain an updated list of the recommended resources and activities included in this chapter. We will also collect tutorials for the stages introduced in this section.

Stage 1: Situate Our *Type of Content*

In the preface of their book *Technical Writing 101*, Alan Pringle and Sarah O'Keefe (2009) jokingly warn their readers that a career in technical communication will be challenging and fast-paced, but "discussing your work at a party will be quite a conversation stopper" (p. 19). Not much has changed since they wrote that preface, and we still have to explain our profession to people outside our discourse community. Adding the word "content" to the work of technical communicators requires even deeper explanations, as the work that we do is not necessarily related to the main type of product that comes to mind when people think or talk about content. Today, primary collective mental images of *content* can be an Instagram food post, an early 2025 TikTok dance challenge, or a YouTube prank video. It can take several steps up the ladder of explaining "web page content" to arrive at the type of—mostly written and structured—computable content that technical communicators work with.

Content, therefore, has strong connotations of ephemeral, entertaining, or even juvenile in the collective, accepted definition, and our profession has to establish differences between our work and that broader interpretation of the term. A good starting point to address these differences in technical communication curricula is to step out from our neighborhood of technical content specialists and read a broader treatise of the term. A short and effective reading that we have assigned in some of our courses is *Content*, by Kate Eichhorn (2022). She acknowledges that "nearly everything that circulates digitally came to be known as content" (p. 5) and connects the term intrinsically to the industry that supports it. Eichhorn adds that content "sometimes conveys information, tells a story, or entertains, but it doesn't need to do any of these things to circulate effectively as content" (p. 15).

Moving from a content-is-everything definition closer to *our* type of content, *Content Strategy*, by Rahel Bailie and Noz Urbina (2013), is still an unrivaled text. The authors start from a perspective that portrays content as ubiquitous in a technology-enabled world, but then they focus on the concept of copy as a gateway to *our* type of content. Copy, in their definition, "is the message. It's what content consumers read; it conveys meaning" (p. 131). They then propose the following two basic genres of copy used in business:

1 Persuasive copy, which tries to convince readers to do a certain thing in a call to action, with formats that include news releases, marketing products, and ads.
2 Enabling copy, which helps readers to complete a process or task with "how to" information, with formats that include procedures and instructions (Bailie & Urbina, 2013).

Carlos worked with Rahel Bailie to summarize the following definitions related to content and copy:

- Copy = editorial; texts: meant for human comprehension
- Metadata = technical; schemas: meant for machine "understanding"
- Content = editorial + technical (Bailie & Evia, 2024; Evia, 2024).

As a result, the combination of editorial and technical creates the kind of content that we are mainly talking about in the content-focused facet of technical communication and that should be introduced in this stage.

Resources

Additional explorations of content and technical communication at this introductory stage include the following:

- Rebekka and Tatiana Batova (Andersen & Batova, 2015) explored and synthesized several definitions of content in the article "The Current State of Component Content Management: An Integrative Literature Review."
- *Intelligent Content: A Primer,* by Rockley, Cooper, and Abel (2015). Another short, influential book that presents intelligent content as structurally rich (content is structured in a way that computers can understand) and semantically categorized (content has tags and metadata that give it meaning beyond the copy).
- The article "When Writing Becomes Content," by Lisa Dush (2015), which is particularly relevant for faculty and students coming to content

from a tradition in writing studies. Dush separates content from other types of writing by describing it as conditional, computable, networked, and commodified.

- The afterword "Beyond Management: Understanding the Many Forces that Shape Content Today" (Evia & Andersen, 2020) that we wrote for *Teaching Content Management in Technical and Professional Communication* provides an overview of the forces shaping what we describe as the discipline of content, and it focuses on structured authoring as its key component.

Activities

When we have addressed this stage in our college courses, we don't expect that most of our students will move to more advanced stages of technical content. Carlos, for example, teaches about different types of content to students majoring in Scientific and Technical Communication, but also to students in the broader Communication degree at Virginia Tech. Students in the latter group frequently take their introductory knowledge of content to conversations related to public relations or journalism in more advanced courses, and they most likely won't become technical content specialists. Rebekka faces a similar audience when teaching about content to students minoring in Professional Writing or students taking her user documentation course to complete a general education requirement.

Therefore, an activity that benefits students at this stage, regardless of their major or future interactions with technical content, is an introduction to the use of basic semantic styles in word processing. The use of styles for headings and subheadings in Microsoft Word, Google Docs, or similar applications presents some early benefits of structured authoring. Students start separating content from presentation and can see their work's sections as chunks clearly separated from each other. Furthermore, styles allow dynamic creation and updates of tables of contents, and changes to fonts and colors in text marked up with styles are automatically reflected in whole documents. In Chapter 3, we cited some sources connecting proper use of styles for headings and subheadings to improved content for AI large language models.

Working in word processing apps using basic or pre-built semantic styles, students at this stage can work on the following assignments:

- Compose an essay exploring different types of content and copy. Expand Bailie and Urbina's taxonomy of persuasive and enabling content to subcategories or other types used in web pages and social media (e.g., entertaining content, marketing content, multimedia content).

- Write the script for a multimedia project combining different types of content. How would an Instagram reel, a fictional podcast episode, or YouTube video present a taxonomy of content? Based on the scope of a specific course, students can use their structured script to actually create the multimedia product.
- Combine written, structured scripts with multimedia content to enhance accessibility in the forms of captions, alternative text, or parallel descriptions.

Competencies

In this stage focused on situating our type of content, students begin to develop the following conceptual skills in the competencies framework for content professionals that we presented in Chapter 6 (refer to Figure 6.1):

- *Systems knowledge and thinking*—through readings and discussions, students come to understand the forces driving the need to treat and manage content as computable data, and they gain insight into how organizations are doing this through adoption of structured authoring workflows; they begin to develop a big-picture view of the processes, workflows, and life cycles that shape content activities and roles and the nature of content itself.
- *Abstraction*—students begin to discern patterns and approach problems in layers; they gain familiarity with the concepts of content analysis, information typing, and content modeling.
- *Rhetorical proficiency*—students tailor communications by thinking about how content chunks might interconnect to serve the needs of different audiences for different purposes; they begin to understand the what, how, and why of structured authoring and the role of user and task analysis in planning content types and deliverables.
- *Experimentation*—students explore possibilities for chunking and structuring content and for combining content chunks and assets in different ways; they adopt a learning and problem-solving mindset.

Stage 2: Introduce Content Strategy

Once some boundaries and definitions for different types of content have been established in Stage 1, students should become familiar with the content life cycle. There are many versions of content life cycles, and we are not presenting the one proposed by Bailie and Urbina as infalible, but it has worked well in our courses. At this stage, students probably won't create content that goes to all four quadrants of the life cycle, but they should be aware of (and prepared for) advanced stages even when they are starting to explore the *Analyze* quadrant.

At this stage, students should concentrate on the disciplinary area of content strategy, which will be beneficial even to those who do not end up working with technical content or take future courses that cover more advanced stages. Most approaches to content strategy focus on key processes that include audience analysis, an audit of existing content assets, and the elaboration of a detailed plan to guide the whole content life cycle.

Instructors need to make a decision about the scope of student projects starting at this stage. Many opt for projects that involve real clients in need of a content strategy or a concrete content deliverable (a new website, updates to an existing online knowledge base, a social media campaign, etc.). Some use fictional projects or ask students to work on projects related to their interests or extracurricular activities. We present ideas for resources and activities regardless of type of project in a specific course.

Resources

In Chapter 2, we mentioned that content strategy is the most developed and documented disciplinary area related to content; as a result, there is a broad variety of materials that introduce this disciplinary area and provide templates and ideas for activities and deliverables. Books that we consider as classics, including the mentioned *Content Strategy*, by Bailie and Urbina (2013), and *Content Strategy for the Web*, by Kristina Halvorson and Melissa Rach (2012), are now a bit dated for an introductory course. We have used the following texts that were well received by undergraduate and graduate students:

- *The Content Advantage* (published as *Clout* in previous editions) by Colleen Jones (2019). Practical, accessible, and effective, this book has tips and ideas applicable to content strategy for all types of digital products.
- *The Content Strategy Toolkit*, by Meghan Casey (2023). A graduate student in a course taught by Carlos described this book as "a spreadsheet for every occasion." Impressive collection of templates and formats to present content strategy as a serious business project instead of an afterthought.
- *Content Strategy: A How-to Guide*, by Guiseppe Getto, Jack Labriola, and Sheryl Ruszkiewicz (2023). Particularly related to technical communication, this resource includes detailed lists of books, articles, and tools usable during the process of developing and implementing a content strategy.

Activities

The three books that we recommend for this stage include detailed activities and templates for planning and conducting a content strategy. Regardless of book or approach, we recommend focusing on audience analysis (emphasizing the multiple layers of diversity in audiences), organizational goals and capabilities (including budget and human resources available), a content audit spreadsheet, and a shareable plan to present as the main deliverable of a content strategy. To continue the adoption of structured authoring, reports and deliverables at this stage should use semantic headings and subheadings in word processing applications.

In Chapter 3, and particularly in Table 3.2, we summarized ideas and opinions from experts to provide an overview of emerging use cases for AI use in the content life cycle. We are aware that the adoption or rejection of generative AI tools for teaching and learning varies according to institutional barriers and limitations and is even affected by the preferences or beliefs of individual instructors. This section does not want to push for the use of AI tools without thinking about consequences and larger considerations for specific cases. However, in recent industry events related to technical content, including editions of the LavaCon conference on content strategy and content operations as well as conferences organized by the Center for Information-Development Management (CIDM), the conversations in formal presentations and informal hallway chats have been about how to manage AI tools in technical communication and not about considering adoption or rejection of these tools. Therefore, we include references to AI tools that some readers can ignore based on their context and preferences.

Taking into consideration the need to introduce generative AI tools in college curricula with disclaimers about ethical use of these tools and the separation of AI-generated text and students' original work, generative AI tools can be helpful in this stage. We have used Microsoft's Copilot and OpenAI's ChatGPT as role-playing tools when conducting audience analysis. With proper prompting, students can use an AI model as a user persona based on the recommendations from a specific book included previously under the Resources section.

At this stage, we introduce students to Rahel Bailie's version of the content life cycle. Even in those courses that work as a standalone exploration of the discipline of content, the life cycle is a helpful resource to foster a collaborative mindset and systems thinking. An additional activity that has been helpful primarily with English students is to compose an essay contrasting the quadrants of Bailie's content life cycle with the rhetorical canons. This activity helps us situate content as a direct heir to rhetoric and communication, and not just an ephemeral process involving digital devices and audiences with short attention spans.

Competencies

In this stage focused on introducing content strategy, students build on the foundations of content established in the first stage, beginning to develop and further developing the following conceptual and practical skills in the competencies framework for content professionals (refer to Figure 6.1):

- *Systems knowledge and thinking*—students become familiar with the concept of a content life cycle and the activities associated with the *Analyze* quadrant; drawing on their understanding of business strategy and processes, user journeys, and product life cycles and development processes, they gain practice developing a shareable plan to guide the content life cycle.
- *Rhetorical proficiency*—students analyze different audiences who have a stake in the content life cycle, from business leaders and members of cross-disciplinary teams to end users, and they develop a business case for their proposed plan of action for content; their knowledge of structured authoring informs their plan.
- *Abstraction*—students draw on audience analysis results and research to discern content use and needs patterns and approach the development of a content strategy in layers; they conduct a content audit and analysis, and they engage in content modeling with reuse needs in mind.
- *Research*—students conduct research to solve problems; they talk with different content stakeholders; examine business goals, capabilities, needs, and existing processes; and analyze and interpret collected data for the purpose of developing a shareable plan to guide the content life cycle.
- *Social proficiency*—students engage in varied collaborative analysis and research activities necessary to develop a shareable plan for the content life cycle; collaborative activities can offer students opportunities to develop interpersonal, intercultural, global, social justice-oriented, and listening skills as well as opportunities to learn about intersections between Agile development practices and technical content.
- *Technological literacy*—in developing a content strategy, students gain proficiency in using project management tools, data analysis tools (including generative AI), and the structured authoring features of word processing tools; they also ideally gain familiarity with editing app and build system options; content management options, such as content management systems or source control systems; publishing processor options; and product development processes.

Stage 3: Focus on Content Creation and Management

This is the stage that tends to keep some colleagues in academia from incorporating activities and courses related to treating technical content as computable data and a valuable asset. For many years, this was the stage that involved a certain degree of tinkering with computer code and a combination of applications involving the command-line interface and many angle brackets. In 2025, technical content can still be created through those lower levels of abstraction workflows, but we are fortunate to have applications now with friendly user interfaces and all-included processes that lower the barrier for students and instructors in the humanities and liberal arts to contribute to content design, creation, and management.

Technical content creation workflows that move from the basic structured authoring allowed by the use of styles in word processing tools are not linear and immediate like writing on a piece of paper or typing words that will only be presented in one way (e.g., Arial bold 12pt in a memorandum to be printed). Work created in this stage should be grounded on the separation of content from format that we discussed in Chapter 3 as foundation to our type of content.

An intermediate action to advance from styles in word processing to truly structured content is the introduction of HTML in courses and activities. In the past, we were guilty of focusing too much on the process of manually writing HTML (and sometimes even CSS) in technical content curricular components. Although we still believe that being fluent and comfortable with HTML is part of the structured authoring competencies, we are also aware that no technical content professionals are hired nowadays to mainly work swimming in hundreds of lines of HTML. Therefore, we now dedicate only a couple of lessons, or a course module in some occasions, to hand-coded HTML in a text editor. Unless a course or assignment is specifically about creating websites as a final and only deliverable, we recommend avoiding web-building tools that focus on presentation and keep content tightly coupled with one template (Wix and Weebly are examples of this type of tool in 2025). Likewise, some popular web content management systems (WCMSs) like WordPress provide a relatively easy way to create and maintain a website, but (unless content engineers enable and use a backdoor of APIs) those are solutions for presentational content that won't work for technical content that can be repurposed and adapted for different audiences and deliverables.

In *Long Life Learning*, Michelle Weise (2021) talks about skill shapes as a way to understand contextual (primarily regional) workforce needs. She explains that one role can emphasize different skills "based on whether an employer is seeking to fill that role in Wichita, Seattle, or St. Louis" (Weise, 2021, p. 134). Thus, the shape of a technical content creator is hyperlocal and requires different skills. In some markets, HTML can suffice as a language to design and manage content. In other markets, primarily those connected to software development, there is a demand for the skill of proficiency with languages like reStructuredText or AsciiDoc, or a more detailed docs-as-code approach. Teaching every possible markup or scripting language for content doesn't seem like a practical solution, and instructors should make curricular decisions based on the skill shapes determined by their regional needs. Over a couple of decades, however, we have identified proficiency with XML and DITA as a transferable skill for technical content development and management across skill shapes. We posit that starting with DITA XML is more generalizable to many skill shapes than starting in a lightweight language and then going to stricter content structures. Furthermore, the tools and workflows commonly associated with DITA truly support the development of content as computable data and a valuable asset. A parallel, also highly transferable skill is proficiency with Markdown, particularly when connected to DITA XML in Lightweight DITA workflows. Since 2010, the syllabi of our content-related courses have actually put Markdown in the spotlight (replacing raw HTML code) as the main way to generate web-aimed content.

This stage also includes managing content assets. In Chapter 2, we wrote that in technical communication academic circles, we have tended to cover everything related to technical content with the banner of content management. When many of our colleagues teach courses or modules that aim to address this process, their most used tool is a WCMS. For the skill shape of a content creator with a primary or only focus on websites, a WCMS like WordPress would be an appropriate tool. However, content management is a complex component of the discipline of content that cannot be reduced to one single tool. Regardless of skill shape or scope of delivery formats or channels, content assets are more flexible, kinetic, and available for reuse and filtering when they are not bound to a specific deliverable. Content assets inside a WCMS can be products of well-executed content strategy, design, and development; nevertheless, they can only be consumed by end users as a single website output tied to a specific design and without customization for unique audiences, products, or contexts.

For example, content for the fictional RobotVac company that we introduced in Chapter 2 can be stored in a WCMS that delivers it to users as one single website. Users with the advanced model *Ya'ab-u-xúuch* will still need to see instructions that include introductory sections (irrelevant to users of advanced models) for all existing models. And any RobotVac user taking their cleaner on a camping trip to a location without Internet access would have to print instructions from the website with the fonts and theme selected for the web output. Content professionals at RobotVac would need to perform hours of copying and pasting (with potential consequences of redundancy and more content assets to update in the future) to create deliverables addressing the diversity of their products, audiences, and usage scenarios. Moving content from the WCMS to a repository that performs as a single source of truth would allow the customized deliverable of instructions with tasks connected to specific models, and it can also enable the automatic production of deliverables formatted for print, web, mobile, and other Internet of Things channels.

In Chapter 2, we briefly mentioned single sourcing (or developing a single source of truth, as many consulting firms call it), which can be defined as the practice of "creating content once, planning for its reuse in multiple places, contexts, and output channels" (White, 2016, p. 56). Single sourcing has been heralded by information management professionals as "the path to liberating a content offering for personalized, customized, or automated expression" (MacIntyre, 2024, p. 93) and has been part of the technical content specialist skill shape in many regions and industries globally. Recently, single sourcing has even been identified as a solution to minimize the risk of hallucinations and mistakes when training large AI language models ("Why your company is struggling to scale up generative AI," 2024).

But the road to the promised land of reusable and multichannel content assets is not that simple, and computer code is not the main obstacle here. Hyperlocal and contextual skill shapes determine the appropriate tool stack for a given industry and specific workflow. Most enterprise-level content management systems for these capabilities are prohibitively expensive for classroom assignments, and some are customized for skill shapes that respond to unique, hard-to-generalize contexts. Some vendors provide free or reduced-cost academic licenses for classroom usage, and there is undeniable benefit to exposing students to these systems when possible. Regardless of availability of more specialized tools, the skill that we propose here as transferable is understanding how to store and manage content assets in a GitHub repository.

Stephen J. Quigley et al. (2022) explain that "GitHub provides server space for storing different versions of software and integrated Git-based tools that developers use to track changes in documentation, thus ensuring that one individual's contributions do not overwrite another's" (p. 247). Although GitHub's primary function is as a version control system, and not necessarily as a CMS, the popularity of content-related connectors to GitHub through Application Programming Interfaces (APIs) makes it a good choice for beginners in technical content environments, and even organizations with complex content needs, to develop a single source of truth. In Chapter 3, as a matter of fact, we reported that 50% of managers who responded to a survey from the research study supporting this book managed their content assets through a version control system instead of a CMS.

GitHub also includes features like GitHub Pages (public webpages hosted and published through GitHub) and GitHub Actions (a continuous integration and continuous delivery platform for automating building and deployment of content deliverables) that have simplified the adoption of multichannel content management workflows in recent years.

Resources

Before we recommend any books or articles related to content creation and management, we need to introduce the Structured Authoring Committee of the Association for Computing Machinery's Special Interest Group (SIG) in Design of Communication (ACM SIGDOC). This committee was founded by experienced content specialist Stan Doherty as an evolution of the defunct DITA Adoption Technical Committee at the Organization for the Advancement of Structured Information Standards. The committee is composed of industry professionals and educators with expertise in structured authoring and is working to develop free resources to support educators who want to teach structured authoring at the assignment, course, or program level.[2] In addition to developing an industry speaker directory and a repository of sample structured content, the committee develops and maintains curricular units that include various instructional modules—from "Teaching Structured Authoring in a DITA-based Environment" to "What is Structured Content?" and "What is GitHub?" The committee also maintains a list of software vendors who have contributed academic licenses for student projects. We have been involved with this committee since its creation, and we have included many of its resources in this book's companion GitHub repository.

Some foundational books and helpful resources in this stage include the following:

- *Developing Quality Technical Information, 3rd Ed.*, by Michelle Carey et al. (2014). The latest edition of this classic has been described by some of our past students as career-defining and "as entertaining as a good novel."
- *Introduction to DITA*, by JoAnn Hackos (2011), which is still relevant although a bit dated in its examples.
- *Plain Language and Ethical Action*, by Russell Willerton (2015). Come for the plain language primer, but stay for the unrivaled overview of ethics in professional and technical communication.

- *Strategic Content Design*, by Erica Jorgensen (2023). Aimed at professionals with skill shapes that do not necessarily focus on technical communication but are very informative and practical.
- *Communicating Project Management*, by Ben Lauren (2018). More than just focusing on Agile as a methodology, Lauren analyzes its pros and cons and contrasts it to other project management approaches—all in the context of technical communication workflows.
- *Docs-as-Ecosystem*, by Alejandra Quetzalli (2023). For regions where Markdown and software documentation are required skills. The author expands the model of docs-as-code to focus on a community-centered approach to documentation development.
- *Markdown Tutorial*,[3] a web resource maintained by Garen Torikian with an interactive walkthrough that we have used for years when introducing this lightweight shorthand language for creating digital content.
- *Learning DITA*,[4] a website from Scriptorium that offers free DITA tutorials. It is a solid resource for novice-to-advanced users who need to learn about DITA and its content structuring and publishing capabilities.
- The YouTube playlist "GitHub for Beginners," from the official GitHub account. Technical content students can skip the "Beginner Git Command" video. Kedasha Kerr's explanations are truly appropriate for beginners.

And we also recommend some of our work:

- The article "Teaching Structured Authoring and DITA through Rhetorical and Computational Thinking" (Evia, Sharp, & Pérez-Quiñones, 2015) includes the foundations of our approach to teaching concepts and activities from this stage.
- In *Creating Intelligent Content with Lightweight DITA* (Evia, 2018), Carlos guides readers through the layers of abstraction involved in the process of planning, creating, and publishing content with the DITA standard and its proposed Lightweight DITA variant, which allows the combination of Markdown, HTML, and XML in repositories to be used in DITA-based content workflows.
- The chapter "Operationalizing Content Creation," by Rahel Bailie and Carlos (Bailie & Evia, 2024), has a primary intended audience of authors who need to navigate the life cycle of technical content without previous experience. The chapter acknowledges that not all content operations are (or need to be) based on complex tool stacks and advanced levels of reuse and single sourcing, and it offers ideas for different levels of robustness required in workflows.

Activities

Keeping activities from this stage ("Focus on Content Creation and Management") disconnected from those in Stage 4 ("Publish or Serve") does not look like a logical decision, but in some contexts, it is necessary. The separation of content from presentation that enables multichannel content publishing and single sourcing sometimes means that authors will not see an immediate "complete" publication of their work. A technical content component (or topic in DITA terminology) with essential steps for all models of RobotVac machines, for example, will appear in several printed manuals, websites, and a mobile app. The human author(s) who wrote those essential steps won't see those deliverables as soon as a given version of the content is committed to the company's repository. Additionally, that same content component could be served to databases for vendors across the globe, and the author simply won't know where their words are being displayed and consumed (or which fonts and background colors will be connected to their content). In this stage, activities should focus on the implementation of the content strategy to produce and manage content assets, but not necessarily in the publication or final presentation of deliverables.

- An activity that introduces students to the concepts of maintaining a centralized repository of content assets and to writing in a topic-based workflow (as opposed to writing full book-length deliverables for one single output) is the creation of a collection of items that requires one content file for each entry. The files can be created in HTML or Markdown depending on the skill shape required in a region or sector. For an updated list of recommended editors and tools, tutorials, and sample collection entries, please refer to this book's companion GitHub repository.
    - Each entry in the collection (in previous courses, we have assigned projects with collections of movies, popular songs, household gadgets, apps on a mobile phone, and even a list of friends or contacts) should be in an individual computer file and follow a structural template. A classroom activity should include content modeling to create the consistent headings and subheadings that will structure every entry in the collection. Students can draft prompts for generative AI models asking for the structure or skeleton of HTML files (even include requests about CSS for presentation), but in our courses, we emphasize that the prompts must mention that content will be generated by the students.
    - Students should create a GitHub repository to store their content assets/collection items. HTML-based collections would be

immediately rendered as user-ready deliverables when seen through a web browser; Markdown-based collections need additional processing that we will address in Stage 4, but GitHub will provide a preview hiding the code.

- A more advanced version of the collection assignment is discussed by Carlos and his co-authors in "Teaching Structured Authoring and DITA through Rhetorical and Computational Thinking" (Evia, Sharp, & Pérez-Quiñones, 2015). This version introduces metadata components to create processing conditions that will filter content components according to the needs of specific audiences. The assignment asks for a collection of DITA topics providing breakfast recipes for a vegan roommate and a carnivore relative. The key is to avoid duplicating recipes and instead use metadata to mark paragraphs, steps, list items, and images with options for both audiences. Students should document their use of metadata and content types from the DITA standard to create a taxonomy, which will be particularly helpful in team projects. We have also taught a different version of this assignment, in which students have to remix a story and include, in the same collection of topics, content for children and adults that could be filtered in the next stage. Figure 7.2 includes an excerpt from the assignment description, which uses "The Story of the Three Bears" as an example. Students can use generative AI tools and prompt for valid DITA code to then populate with their original content. For more specific tutorials and software recommendations, please refer to this book's companion GitHub repository.

Competencies

In this stage focused on content creation and management, students build on the competencies that they developed in the first two stages, and they begin to develop expertise in structured authoring:

- *Systems knowledge and thinking*—through readings, discussions, and activities, students continue to develop a big-picture view of the processes, workflows, and life cycles that shape content activities and roles, and the nature of content itself; drawing on the content strategy that they developed in Stage 2, they create content assets that have the potential to interconnect with other content assets, workflows, and systems.
- *Rhetorical proficiency*—students learn the what, why, and how of separating content from format, the foundation to technical content, and practice creating content assets that are modular, flexible, kinetic, and reusable; they apply plain language and minimalism principles to their

FIGURE 7.2 Sample DITA Topic with an Excerpt of "The Story of the Three Bears," by Flora Annie Steel (in the public domain and retrieved from Project Gutenberg). This sample, created in the Oxygen Content Fusion authoring platform, shows a topic with phrase (<ph>) elements providing structure and metadata for different types of audiences ("child" and "adult" in this case). Content for both audiences resides in the same source file and will be filtered out when processing at a later stage. Conditional phrases are used to modify the wording based on the audience attribute used as metadata, with the phrase *said* appearing for a child audience and *roared* appearing for an adult audience. *This eBook is for the use of anyone anywhere at no cost and with almost no restrictions whatsoever. You may copy it, give it away or reuse it under the terms of the Project Gutenberg License included with this eBook or online at www.gutenberg.org.*

writing, and they write with user goals as well as reuse and multichannel publishing goals in mind.

- *Abstraction*—students practice approaching content creation and management in layers of abstraction, from creating format-free content assets to using metadata to create processing conditions that will filter content components based on the needs of diverse and complex human and machine audiences. They focus on accessibility and inclusion in language and multimedia elements.
- *Experimentation*—students experiment with structured authoring approaches, standards, and tools as well as generative AI models that

can suggest structures or skeletons for HTML, DITA, and other file formats; they adopt a learning and problem-solving mindset.

- *Technological literacy*—students gain proficiency in applying methodologies and using tools that support structured authoring and management of structured content, including editing apps, build processors, and source control systems (and maybe even CCMSs), and they learn strategies for using AI tools to answer questions, offer suggestions, validate content, and generate standards-compliant outputs; students also develop fluency in reading and writing HTML and DITA XML and, depending on course and assignment foci, fluency in reading and writing CSS and Markdown.

Stage 4: Deliver and Maintain

This stage is without a doubt the most difficult for instructors and students who are primarily focusing on the disciplinary areas of content strategy and/or content design or creation. This stage is an early foray into the work of content engineers, and although it is an advanced level, we have deemed it necessary in our courses. Without it, students do not get that "Christmas morning effect" of seeing their content assets become formatted and presentable information deliverables.

In the previous stage, students created GitHub repositories with content assets authored in HTML, Markdown, or DITA XML. However, those raw components have not yet produced deliverables that could be shared with end users.

In Chapter 3, we mentioned the DITA-OT[5] as the primary engine for transforming DITA (and even Markdown) content assets into a variety of user deliverables. In the past, we had to introduce our students to the DITA-OT so they could see their work become a website, PDF, or EPUB file that they could distribute to human beings. Nowadays, we have resources that hide the complexities of working directly with the DITA-OT in a command-line interface. We could say the same about the powerful Pandoc[6] document converter, which integrates Markdown files to a multichannel publishing environment, but its tool stack requires a level of technical abstraction that can intimidate entry-level content creators.

Resources and Activities

Resources and activities in this stage are connected to those presented in Stage 3. We re-emphasize that the ACM SIGDOC Structured Authoring Committee maintains a list of software vendors who have contributed academic licenses for student projects and we reference those in this book's

GitHub repository, where we also include tutorials and sample workflows for creating DITA content and storing it in GitHub repositories.

If an instructor has access to DITA processing software, the application most likely includes a transform or publish feature that assembles DITA topics and maps into deliverables for specific audiences. GitHub Actions also enable processing inside repositories, and a workflow can—for example—invoke the DITA-OT automatically in a process of continuous integration and development to build and serve website and PDF deliverables. Based on the work of some content engineers who contribute to the DITA-OT, we have included a template (and instructions for using it) with DITA-OT GitHub Actions enabled in this book's companion repository. We also include in the companion GitHub repository a template for publishing Markdown files as a website using GitHub Pages, and a repository that uses GitHub Actions to call the Pandoc document converter and continuously build PDF and EPUB deliverables from a collection of Markdown files. We include, as well, a tutorial for publishing DITA files (topics and maps) as a website using GitHub Pages.

Competencies

In this stage focused on delivering and maintaining content, which is closely aligned with the *Maintain* quadrant in the content life cycle, students develop more advanced conceptual and practical skills in the following areas:

- *Systems knowledge and thinking*—drawing on the content strategy that they developed in Stage 2, students define the life cycle for their content and implement a workflow that accounts for complex and diverse user journeys; they may also consider ways in which their content might be repurposed and reused in other content workflows.
- *Abstraction*—in planning for reuse and interactions with other content workflows, students might develop and implement a metadata and taxonomy strategy.
- *Experimentation*—students have opportunities to explore different tools and systems and use them to problem-solve.
- *Technological literacy*—students develop more advanced proficiency with editing apps and build processors as well as source control or content management systems, and they gain experience working with publishing environments; given growing integrations of AI into content tools and systems, students continue to develop AI and code literacy—if the instructor decides to focus on source code.

Before, during, or after integrating one or more stages of content work into curricula, technical communication educators can focus on innovation through industry outreach, interdisciplinarity, and professional development. We present ideas for innovating curricula in the next section. Building connections with industry partners, particularly in the form of advisory boards and internship and mentorship programs, and through prioritizing interdisciplinarity in curricula, educators can create opportunities for students to develop additional knowledge and skills that complement those developed in the four stages. Additional knowledge and skills, as represented in the competencies framework for content professionals (refer to Figure 6.1), might include hardware, software, and domain knowledge, as well as social proficiency and research skills developed through experiential learning and team-based projects.

Enhancing Curricular Pathways

In this section, we share the advice that both industry and academic leaders had for technical communication educators on how they could keep curricula aligned with the needs of the content-focused facet of the field. These recommendations apply not only to curricular pathways within technical communication and related degree programs (e.g., emphases, specializations) but also to minors, academic certificates, and individual courses. They apply, as well, to continuing education and professional certificate programs. We also offer ideas for professional development, gleaned from our own experiences and from the conversations we had with industry and academic leaders.

In our interviews, we asked leaders two main questions:

- What do you see as implications of changing roles and hiring needs for education in technical communication (undergraduate degree programs, minors, certificates)?
- What advice would you give administrators of technical communication programs to make sure their programs were meeting the needs of the profession?

The biggest implication of changing roles and hiring needs, and the number one recommendation for educators, was to emphasize structured authoring throughout curricula. We describe ways programs might do this in the "Building Curricular Pathways" section. Here, we focus on other key recommendations for educators. These include, among others, making outreach to industry partners a priority, valuing curricular agility, and encouraging minors and electives in business and leadership and in technology-focused disciplines. Preparing students to work in business environments through a focus on business processes, Agile methods and approaches, and product life cycles—ideally connected to project-based learning and cross-disciplinary team-based learning—was a key theme. Industry leaders, in particular, saw these recommendations as a way for programs to help students develop the competencies that they identified as important for content professionals and for career mobility—that is, for securing entry-level positions and then moving into more advanced, strategic positions that afford them opportunities to exert influence on their organizations.

In what follows, we discuss how programs can keep curricular pathways for content professionals aligned with industry practices through industry outreach and partnerships, interdisciplinary alliances and curricula, and professional development.

Industry Outreach and Partnerships

The need to engage in more industry outreach and build robust, long-term partnerships with peers working in industry was emphasized more than any other topic in response to our questions about implications for technical communication education. Industry and academic leaders agreed, across the board, that academic programs, experienced content professionals, and professional organizations need to work together to bridge the education divide in the content-focused facet of the field. For us, the clear takeaway from our conversations with leaders was that more cross-talk and more frequent collaboration is the best way for academic programs to be well-positioned to prepare the next generation of content professionals.

Industry leaders had much to say about the value of building partnerships with academic programs and participating in initiatives and activities that support students' development of competencies and engagement in experiential learning. We offer representative comments below:

> I think the programs have to have an outward-bound focus or awareness. They don't necessarily need to build their own programs to get ready for supplying all these people.... They don't have to build those programs; they just need to build the partnerships so they understand the problem and then we can help them. They can't build it ex nihilo; they have to get to understand it and start to cross-fertilize, get people to visit, chat. I don't think there's any reticence, at least that I've ever seen, from people in the industry to want to help out. It's just for some reason, we've just somehow lost that outreach. I would say partnerships is the missing ingredient. Figuring out, from either side or both sides simultaneously, how do you build these relationships? How do we talk to one another? How do we understand what's going on? I think there

> are enough smart people around that if everybody had more insight into the challenges you have as a professor and the challenges I have as somebody in the industry, we'll figure it out. But right now, we're, I'd say, flying blind. We're kind of flying blind.
>
> —*Senior documentation manager (interviewee)*

> In the early 90s, several schools developed what they called an IS program, which was a combination of business school and computer science. They were not developing computer scientists of the future; they were developing computer savvy business people. One of the things they got right, that computer science still has not gotten right, and that tech comm does not do, is the importance of your network and why you do it, why you socialize, why you build it, why you get involved, why you keep up with people. That whole thing, that's a business school basic that nobody else does. They did it right for the IS school, but computer science never picked that up. I think that's one of the things that's super missing from tech comm programs.
>
> —*CEO and founder of a content solutions consultancy (interviewee)*

> ...I talked to some of the conference organizers, and I asked people, 'What's going on here, we need young people, we need energy." We talked, and we said, 'Maybe we should actually have a focus on... I guess on reaching out to academic institutions and saying, how do we get people into the industry, how do we attract young talent?' That's when I came back and made a conscious decision that I was going to reach out to local universities and see if we could hire some interns.
>
> —*Senior director of user assistance (interviewee)*

We heard concerns from several industry leaders about both educational programs and hiring managers not talking to each other enough. The general sentiment was that neither side tends to be aware of the challenges the other side is facing nor the problems they are trying to solve. A key takeaway from these conversations was the message that many industry peers want to help—they want to engage with educational programs and students and forge mutually beneficial partnerships.

Academic leaders also had much to say about the value of building partnerships with local industry to identify employer needs, establish advisory boards, build repositories of guest speakers and workshop presenters, develop mentorship and job shadowing programs, establish internship and co-op programs, and develop experiential learning opportunities such as service-learning and client projects. The comments below speak to the strategies that these leaders have employed to keep program curricula aligned with professional practice.

> If you don't have the expertise in your faculty to integrate all of the kind of specifics we're talking about, I think the first step might be to establish relationships with industry locally. And with the job markets.
>
> —*Professor of technical and scientific communication and director of professional writing program (interviewee)*

> … my predecessor had set [my program] up with an industry advisory board, and so they advised on the program content at the time. We have kept up those connections, and what has been most useful about that is just knowing that we can have our finger on the pulse, that if something is coming up that we don't react to everything that comes up in industry. Students always appreciate having someone come in and talk about their work as well because it gives them an understanding of what it's like in the real world.
>
> The other thing that we have is really good contact with alumni and that keeps us in touch. And those alumni are generally willing to come and do workshops or talk about something new that's going on in their company. And occasionally will you give us projects that our students can get involved in. So those alumni contacts are essential to us now. And then I think they are really, really important.
>
> —*Professor and head of school of technical communication (interviewee)*

> We're working to respond in concrete ways throughout the curriculum to the needs that our advisory board members have highlighted.
>
> —*Professor of writing studies and former director of technical writing and communication program (interviewee)*

> For our technical communication program, we're thinking about making those connections. We ask them [companies in the area], what are we missing? What do you need? That's one part of our curriculum redesign that we're considering. Just asking what they need.
>
> — *Professor of English and faculty member in graduate technical communication program (interviewee)*

> It's a matter of reaching out to kind of those prospective employers and kind of cutting a bargain with them to achieve this parallelism we talked about. Look, we can provide this with the core structure we have, but

> we can't provide this. Would you be willing to have employees conduct workshops, meet with or talk with our students, provide job shadowing programs, allow for guided internships, even engage in service-learning opportunities that allow our students to acquire the skills or the component we can't provide ourselves? Now by foisting that applications part off onto the employer, you keep your program adaptable to avoid what you pointed out earlier. You don't hire people with skill sets that are dated by the time they get tenure.
>
> —*Professor and endowed chair of technical communication (interviewee)*

For these academic leaders of technical communication programs, keeping their finger on the pulse of industry is key. They do this through outreach and establishing sustained partnerships, primarily through advisory boards. They view partnerships as bridges, where industry partners can make up where academic programs leave off—that is, where programs can lean in on the expertise and experience of industry partners to support curricular needs and where industry partners can learn what kinds of training and mentorship their organizations need to provide. As one leader noted, academic programs should consider how they can transfer some applied learning to industry partners. This is how programs can fill curricular gaps and competencies gaps. This is also how programs can ensure curricular agility, or what one professor and chair of a technical communication department called a "dynamic curriculum." Such a curriculum requires programs to think about the lifespan of individual course offerings and be ready to redevelop courses in response to changing technology and business needs.

Another academic leader, a professor and program director of user experience, agreed, suggesting that curricula must be more agile now:

> We know we have to pivot. We know we have to keep an eye on… the field changes so fast, right? So [programs] need to have enough flexibility in the curriculum to make sure that if we see there is an established trend, or even if it's an emerging trend, we can still have maybe a few courses, some activities, assignments, maybe it's not a course but it's assignments that would target that trend.

Achieving curricular agility also requires educators to examine the local industries that employ program graduates and the kinds of industries that might move into the area that would employ them. Doing so better equips programs to innovate curricula to align with local employment needs. Drawing on the perspectives and experiences of members of a robust advisory board is one tested way programs do this.

As Rebekka (Andersen, 2011, 2014) has argued elsewhere, a reciprocal relationship between academia and industry is critical to the ability of educational programs to keep curricula and programmatic goals and outcomes aligned with professional practice, particularly practice in the content-focused facet of the field. Given sweeping changes in the roles, activities, and outputs of content professionals, Rebekka (Andersen, 2014) called for a praxis-based, collaborative model for education and research. She described how industry needed academia to better prepare content professionals for new roles and practices while academia needed industry to better train and onboard new content professionals and to offer more opportunities for educators and students to learn from content leaders. She argued, "this mutual interest is where productive dialogue between the two groups might begin" (p. 143).

Tracy Bridgeford and Kirk St.Amant's (2015) edited collection, *Academy-Industry Relationships and Partnerships*, articulates several initiatives and ideas for forging reciprocal relationships. The collection was inspired by the 2011 Partnership Preconference at the annual conference of the Council for Programs in Technical and Scientific Communication (CPTSC), which brought together academic and industry professionals to discuss strategies for creating collaborative partnerships that benefit students. Commenting on the value of the collection, St.Amant urged readers to view the initiatives and ideas as just a sample of approaches that programs can take to create and maintain effective academia-industry partnerships (p. 8). Carolyn Rude (2015), who has long called for stronger relationships between academia and industry to secure the field's strength and sustainability, offers the forward to the collection, reminding readers that "Industry depends on academic classrooms to prepare qualified employees, and academia depends on industry's willingness to hire graduates in order to justify academic programs" (p. v).

Ongoing collaboration between academic programs and industry partners will be essential to preparing future content professionals for success in a world where content becomes computable data. Fortunately, outreach and partnerships can take on many forms.

Program-Level Outreach and Partnerships

At the local level, industry outreach and partnerships might take the form of advisory boards, mentorship and job shadowing programs, internship programs and co-operative educational experiences, and guest speakers, workshops, and service learning. In the next sections, we offer ideas for developing partnerships with local industry.

Advisory boards. Advisory boards are perhaps the most effective and productive form of industry outreach and engagement, as they provide an infrastructure for other partnerships and initiatives. Importantly, industry partners who serve on advisory boards can help shape program and curriculum development in the disciplinary areas of content design, content strategy, content engineering, and content operations.

The Technical Communication Advisory Board (TCAB) at the University of Minnesota Twin Cities serves as a model for how programs can develop sustained, mutually beneficial partnerships with local industry that support not only curricular innovation and agility but also student mentorship and experiential learning opportunities. Established in 2013, the 15-member board of business leaders aims to "provide exemplary networking and experiential learning opportunities for students, and to enrich the curriculum and visibility of our programs, students, faculty, and staff" (Duin & Tham, 2018, p. 46). Faculty leaders create opportunities for board members to engage with each other and faculty through regular meetings and roundtable discussions and to engage with students through webinars and workshops, a mentorship program, guest speaker visits, internships, and networking events. One outcome of the regular meetings has been input on course redesign, which has to date helped the BS, certificate, MS, MA, and PhD programs in technical communication keep curricula aligned with industry practice, including modern content development and management practices. Faculty leaders of the advisory board view their partnership with board members as a reciprocal relationship, as evidenced in the board vision statement (2024): "Input, direction, and engagement of TCAB members elevates the presence, relevance and status of our programs while also increasing the effectiveness of our TCAB members and their industries." We recommend referring to Duin and Tham (2018) for a rich description of the history of the board, how its members contribute, and how the TCAB leadership team employs a community-of-practice approach for advisory board engagement.

George Mason University offers another exemplary approach to creating and sustaining an advisory board that is engaged in program and curricular design. Refer to Heidi Lawrence et al. (2023) for a rich description of how faculty in the master's and graduate certificate programs in professional and technical writing created a board comprising representatives from across local industries to provide curricular guidance for their programs. Faculty created the board in response to the need to grow program enrollment and to better prepare students for "meaningful professional and research opportunities" and soon thereafter "conducted a series of focus groups to gain their insight and feedback" (p. 42). Lawrence et al. (2023) describe their research-based approach to engaging board members and offer educators guidance on how they might develop a similar model that centers accountability, emphasizing "collective action, long-term relationship-building, and sustained outreach as mechanisms for opening up spaces and opportunities for PTW students, thus creating stronger and more just PTW programs" (p. 42).

Mentorship programs and job shadowing. Mentorship programs, as well as internship programs, which we discuss next, are two well-known ways to build sustained and engaged partnerships with industry. Mentorship programs provide a way for experienced technical communicators, including leaders in the content discipline, to share their experiences, expertise, and advice with students. Students paired with a mentor working in industry gain not only unique learning opportunities but also networking connections and one-on-one career guidance. If a student who aspires to work in the content-focused facet of the field is paired with an experienced content professional, that student might additionally be given the opportunity to job shadow their mentor and learn about the larger content ecosystem in which they work.

Mentorship programs connected to advisory boards can be relatively easy to establish. The TCAB at the University of Minnesota Twin Cities offers a mentor program that board members can opt into. The faculty leadership team invites board members to participate in the program each fall via an email that asks interested members to report their areas of interest and expertise on a Google Form. Students can apply to participate in the program and are also asked to indicate interests and goals. The leadership team then pairs a board member with one or two students and launches the program in late fall via a TCAB mentor program kickoff meeting where mentors and mentees connect for the first time. The mentor/mentee pairs discuss goals for the program and schedule at least two meetings before the program ends in March. Most pairs meet more than two times, and many schedule job shadowing days.

For educators interested in building a mentorship program, we recommend reviewing Herb Smith's (2015) guidelines for building mentoring programs. The guidelines, among others, include pairing "mentees with mentors who share similar career interests" and requiring mentor/mentee to (1) "meet and communicate frequently to 'cultivate rapport and trust,'" (2) "establish goals and a schedule for the program early on," and (3) connect frequently over the course of at least one full term (pp. 107–108).

Describing one of many benefits of being paired with a mentor, a senior documentation manager whom we interviewed talked about the value of students getting a glimpse into a really messy problem and how content teams are approaching that problem, which might be anything from planning a DITA migration to figuring out how to position writers as integral players in Agile product development processes. He noted,

> We're now all contributing little bits and pieces to something that's actually quite huge and, in some sense, incomprehensible. It's bigger

> than any one of us really understands and can appreciate. I do not know how to expose folks, either at the faculty or at the student level, to that class of problem unless it were in the context of a partnership.

Mentors can offer students rich insights into the kinds of problems that content professionals are grappling with in organizations that have adopted digital content transformation workflows. Many of these problems cannot be reproduced in a classroom environment, and the sheer scope of them can be challenging to explain outside of the context in which they exist.

Internship programs and co-operative educational experiences. Internships are another way that experienced content professionals—in their roles as supervisors and peers—can offer students rich learning opportunities. Internships provide opportunities for students to apply what they are learning in the classroom to workplace situations and to gain hands-on experience with the processes, standards, and tools of the trade. Educators in the field have been discussing the benefits of internship programs for decades, and given the critical role of experiential learning in preparing aspiring content professionals, such programs are more important than ever to include in content-focused curricular pathways.

Internships in organizations that have adopted digital content transformation workflows offer students opportunities to develop and hone the systems knowledge and thinking skills now critical to performing well as a content professional. Likewise, they afford students opportunities to apply—in messy, actual, and low-risk (learning) contexts—the conceptual skills in rhetorical proficiency, abstraction, research, social proficiency, experimentation, and technological literacy that we identified in the competencies framework for content professionals and that programs offering curricular pathways for content professionals need to cultivate in students.

The alumni whom we interviewed who had completed internships all landed full-time jobs as technical communicators within a year of graduating (five of these alumni landed content-focused positions). Many of the alumni were hired for full-time positions by their internship employer after completing their internship. Because applied experiences are so highly valued and because training an intern is a major investment, particularly for organizations with mature or developing content ecosystems, organizations are often motivated to hire strong interns full-time. This internship-to-career pipeline in the content discipline is not a guarantee, but it is certainly a pattern that we identified among the alumni whom we interviewed. And it is a recurring theme in conversations that we have had with industry leaders.

As we described in Chapter 4, while managers of content teams want to see more college graduates entering the profession and would like to be able to support interns as a way to recruit more graduates, many managers are not in a position to do so given resource constraints. At the same time, many of the hiring managers who participated in our research also commented on how they would be more likely to support interns if they had enough know-how to contribute to work activities without extensive training. The amount of time it takes to onboard an intern who does not have experience with structured authoring or tools that support structured authoring workflows is substantial. Undergraduate programs that offer curricular pathways that prepare students as content professionals will be better positioned to work with managers of content teams to develop internship opportunities.

We have also learned from the conversations that we have had with hiring managers over the years that many are unaware of the technical communication degree programs in their geographical region (particularly those working in metropolitan areas). Those managers thus tend not to think about reaching out to administrators and faculty to inquire about building partnerships or establishing internships. A senior director of user assistance at an enterprise software company told the story of going to an industry conference and realizing that the vast majority of folks in the room were her age or older; she asked the conference organizations why no young people were in attendance, and their response was that young people didn't seem interested in the profession. The senior director decided that she was going to personally reach out to local universities and see if she could hire some interns, and after getting approval from management, that is what she did. But her story is an anomaly. Hiring managers in the content discipline, for the most part, are not well practiced in reaching out to local universities. Thus, educators who want to establish internship opportunities designed to support aspiring content professionals should consider reaching out to local industries to gauge interest and raise awareness of their programs. Having an advisory board can certainly help with this effort. Other resources educators can leverage are professional organizations that support content professionals.

Given the critical role of content-focused internships for students wanting to pursue careers in the content discipline, we encourage programs to require students to complete an internship as part of their degree requirements. A designated internship coordinator could work with local industry partners to create opportunities and oversee learning experiences. We acknowledge that many programs, particularly those in remote locations, are not well-positioned to create internship opportunities with organizations that have adopted structured authoring workflows. In these cases, programs might encourage students to do a summer internship with an organization in a different geographical region. Programs might also develop partnerships with campus IT departments, educational technology centers, and/or student disability centers. These entities often have extensive needs for technical documentation and may be open to supporting student interns who can apply their skills in structured authoring and content management to develop new or revamp existing documentation.

For guidance on establishing, coordinating, and sustaining a robust internship program, we recommend reading Herb Smith (2015), Susan Katz (2015), Elisabeth Kramer-Simpson (2018), and Charles Sides (2021).

Guest speakers, workshops, service learning. Building partnerships with experienced content professionals is important not only for developing mentorship and internship programs but also for planning for guest speakers, workshops, and service-learning projects—all activities that afford opportunities to bring industry practices to classroom contexts. Guest speakers are a particularly effective way to introduce students to the different roles and activities of content professionals and the varied environments in which they work. Experienced content professionals who are willing to present workshops are a bonus—workshops on activities that require specialized skills, such as developing a metadata or content reuse strategy or using a docs-as-code approach to structured authoring, can greatly enhance student learning.

An advisory board can be leveraged to support these activities, but educators might need to expand their network to include alumni working as content professionals, members of local professional associations focused on content, and local leaders on LinkedIn who engage in conversations about content. Experienced content professionals who are not local but who are invested in education and interested in recruiting talent to the profession can also be excellent speakers and workshop presenters. Many industry leaders and hiring managers, as well, contribute to freely available podcasts and webinars focused on different aspects of the discipline of content, from ways in which AI is being integrated into content workflows to strategies for unifying technical documentation and product information. Educators who listen to podcasts or webinars of particular interest might follow up with featured speakers to inquire about guest speaking or presenting a workshop.

Outreach to industry peers can also lead to service-learning and client-based projects that give students valuable applied learning experiences and opportunities to further hone competencies such as social proficiency, experimentation, and research skills. Such projects are ideal for cultivating in students a learning mindset and the critical problem-solving skills they need to address messy, ill-defined situations. In the words of one academic leader whom we interviewed, "a good team project with a client… is good because [students] develop that ability to deal with confusing situations that don't have a script written already." Commenting on the value of service-learning and client projects, another academic leader similarly asserted, "This is the place where all the things come together, all those small pieces of what they have been learning—that's what happens in the workplace."

Field-level Outreach and Partnerships

Both the academic and industry sides of the content-focused facet of the field have a lot of work to do to build the bridges necessary to prepare the next generation of content professionals. Over 10 years ago, Rebekka (Andersen, 2014) argued that a praxis-based, collaborative model for technical communication education was necessary if the field was serious about aligning its educational programs with modern content practices. Our professional organizations and individual programs have made some progress for sure, but a more strategic initiative driven by a group of academic and industry stakeholders would go a long way to address the knowledge and resource needs of our programs and faculty.

At the field level, our academic-oriented professional organizations such as the CPTSC and SIGDOC are best positioned to engage in outreach and establish partnerships with industry-oriented professional organizations such as the CIDM.

The field has models of such collaboration. For example, at its annual conference in the years 2011–2014, CPTSC partnered with the Academic Special Interest Group of the defunct Society for Technical Communication. During these years, CPTSC and the SIG successfully brought together academic and industry professionals to discuss the implications of industry trends for educational programs in technical communication. This partnership inspired the collection, *Academy-Industry Relationships and Partnerships*, edited by Bridgeford and St.Amant (2015). ACM SIGDOC, as well, has worked to bridge the divide. The conference committee now encourages practitioner participation through "industry insight" or "experience report" proceeding submissions and presentations. The conference committee also includes an industry liaison. The CIDM, too, has been offering undergraduate and graduate students free registration for its conferences in exchange for minimal volunteer work at its conferences. We find the work of the SIGDOC Structured Authoring Committee, previously mentioned in this chapter, particularly promising.

Partnerships between our professional organizations have the potential to afford more opportunities for each group to expand their knowledge and skill sets, resulting in a positive impact on technical communication practice and training. Rebekka (Andersen, 2014) suggested several possibilities for mutually beneficial collaborations:

> Our academic organizations, such as ATTW and CPTSC, might collaborate, for example, with [content-focused industry-oriented] organizations

> to establish a professional-development exchange program. In this program, the academic organizations could offer workshops and Webinars on topics of interest to industry peers, such as research methods or curriculum development; in exchange, [industry-oriented] organizations could offer workshops and Webinars to academic peers on topics such as topic-based authoring, information-experiences modeling, and multichannel publishing. These academic and industry organizations already offer such professional development opportunities to respective members, so setting up this program might initially be a matter of encouraging participation and overcoming access and time challenges.
>
> *(p. 145)*

We urge our professional organizations to prioritize creating more opportunities for members to engage in productive dialogue, at our respective conferences, via professional development events, via edited collections and special issues of journals, and via discussion boards and social media channels. Currently, industry conferences and workshops offered by industry-oriented professional organizations are cost-prohibitive for educators wanting to engage in dialogue and learn about trends, approaches, and use cases in the content discipline. But given expressed interest by our industry peers to connect with educational programs and to help shape programs and curricula, we see an opportunity to brainstorm with our peer organizations ways in which we could form mutually beneficial partnerships. We also see an opportunity, perhaps through outreach to individual members of our industry-oriented professional organizations, to invite sharing of and even collaboration on onboarding and other training materials designed for new hires. These materials might then be adapted for courses and for faculty development.

In the near future, we hope to convene a group of academic and industry stakeholders to take up the charge of developing a strategic initiative that explores the possibilities discussed above.

Interdisciplinary Alliances and Curricula

As we described in Chapters 3, 4, and 6, hiring managers working in the discipline of content are looking for content professionals who can work and think at a system level and apply technical expertise to both understand complex products and systems and contribute meaningfully to content ecosystems. Students preparing for careers as content professionals, then, would benefit from opportunities to learn about the larger systems that define content activities and the ways in which cross-disciplinary teams and individuals collaborate and interact to achieve respective and shared goals. Given the rise of Agile practices and their impact on the day-to-day work of content professionals, students would also benefit from opportunities to participate in software and product development processes and to engage with engineers and developers to help plan projects and develop service and support documentation. Further, studying business management would help students understand the philosophies, processes, and strategies that drive business decisions and better position them to advocate for end users and fellow workers and regularly practice the kinds of reflection, critique, and research necessary to propose and implement change. In the systems-based contexts in which most content professionals now work, knowledge of business strategy and processes is essential for advocacy work.

We argue that a way for programs to cultivate these new kinds of competencies and offer students opportunities to engage in product development and business practices is by *forming interdisciplinary alliances and integrating interdisciplinarity into curricular pathways.* Technical communication educators cannot possibly teach all of these competencies or offer all of these opportunities, and they do not have to. We encourage educators to embrace the collaborative, interdisciplinary model for technical communication education promoted by Ann Duin, Jason Tham, and Isabel Pedersen (2020). They assert that "collaboration is an imperative disciplinary and workplace assumption that must be taught and practiced" (p. 170), and they encourage programs to not only teach collaboration as a science and practice but also to develop opportunities for students to participate on global virtual teams as well as cross-functional teams that allow for the "cross pollination of knowledge and skills" (p. 174). They also point to Annie Mendenhall's (2013) problematizing of the overemphasis on vertical coherence of curricula and cite her argument, with which we very much agree:

> [W]e need to think vertically, horizontally, and institutionally about how to create courses and curricula. In other words, minors, majors, and graduate programs increase the field's legitimacy by shaping it into a model discipline, but our work might also operate outside the vertical model to engage other disciplines and communities in writing instruction or interdisciplinary programs of study.
>
> *(p. 97)*

Many of the industry and academic leaders whom we interviewed agreed with this sentiment. An information architect and CEO of a communications consultancy commented on the importance of students being prepared to work in business environments and being taught how software development processes are connected to business processes. Because most technical communication programs do not prepare students for entering a business environment, students tend not to be familiar with business processes and or have experience conducting business process analysis. This is an important skill, he said, "because so often what you're doing when you're creating new software is figuring out what's the business process, how can we improve it, and where does software plug into that." In Agile development environments, in particular, writers need to understand how software development processes are connected to business processes.

Similarly, a senior director of user assistance commented on the importance of students being prepared to work in software development environments (where Agile practices shape content practices):

> Think of any application that you use, even a browser—I'm on a computer, but how does that browser get to me? How does somebody decide how it's... What was the concept? What was this idea? How do you go from idea to delivery to consumer, and how does the consumer consume it? That process, from conception to delivery, to even obsolescence, I think is something that everyone should understand. That's something that's definitely missing sometimes in entry-level writers.

A clear takeaway from our conversations with industry leaders was the need for undergraduate programs to form interdisciplinary alliances, to integrate interdisciplinarity into curricula, and to encourage minors in business, computer science, and/or engineering.

Many academic leaders whom we interviewed have long been thinking about interdisciplinary alliances and interdisciplinary curricula. One leader had the following to say:

> I just think it's paramount that technical communicators understand how to work coordinating, cooperating, and collaborating and that... we weave more of this work throughout our curricula. How might we better collaborate...across our institutions more? How do we develop that strong community? I feel very strongly about a connectivist theory...it's not what you know, it's who you know and where you can find the answer when you need it. How might we function more as a collective, cooperative, collaborative body in this field?

A solution, argued another leader, is for programs to become transdisciplinary or interdisciplinary. The key to thinking transdisciplinary or interdisciplinary, the leader suggested, is to not think of the field as belonging to English or communication or any other discipline. It's a field that "draws upon a variety of disciplinary backgrounds and traditions."

Below we describe three strategies programs can consider for forming interdisciplinary alliances and integrating interdisciplinarity into curricular pathways. These strategies complement the knowledge, skills, and experiences students gain through internships and mentorship programs—as well as the knowledge and skills students develop in the core courses comprising a technical communication degree program. Building alliances with other disciplines is a way to fill in curricular gaps and ensure students have opportunities to develop the wide varying competencies identified in the competencies framework for content professionals. Strategies include integrating disciplinary courses into curricula (core and elective), encouraging minors, and prioritizing project-based learning and cross-disciplinary team-based learning, particularly with faculty who teach courses in computer science, engineering, and business.

- **Integrating disciplinarity courses into curricula.** A relatively easy step programs can take to create interdisciplinary curricular pathways that prepare students as content professionals is to include courses from other disciplines as required core or elective courses in the undergraduate major (or minor or certificate). Courses with minimal or no prerequisites from computer science, engineering, library and information science, data science, business management, organizational leadership, and technology management would all be good candidates. Courses focused on programming, artificial intelligence, software and product development processes, business analysis and modeling, web analytics and analysis, project management, data analysis and management, and taxonomy development would be particularly useful.

  To maintain curricular coherence and offer a holistic experience for students, we recommend a capstone course that aims to bring disciplinary threads together. The course would ideally focus on interconnections between the systems, processes, and technologies with which content workflows and content professionals interact. The course might include a team-based service-learning or client project that affords students opportunities to apply their knowledge and skills and reflect on their processes, decisions, and learning journeys.

  We can look at the curriculum of several degree-granting technical communication programs as examples of how disciplinary courses might be integrated. The Bachelor of Science (BS) degree in technical writing at Carnegie Mellon is a particularly robust example. The degree

  > blends coursework in writing, document design, and user experience methods with additional work in computer science, mathematics, and statistics. Students specialize further through elective courses and

professional experiences, earning a concentration in either science and medical communication, or technical communication.

(Department of English, 2025)

Students who pursue the technical communication concentration take three advisor-approved elective courses in management, technology, and/or social issues. The Bachelor of Arts (BA) degree in technical communication and information design at the University of Colorado, Colorado Springs is another notable example. Students take a core of courses focused on varied topics and practices in technical communication and then select an area of concentration that requires them to take three courses in a different discipline. Concentration areas include business administration, cybersecurity management, data analytics, design and media, and programming foundations.

- **Encouraging minors in business, computer science, and/or engineering.** When we asked industry and academic leaders to talk about what they saw as implications of changing roles and hiring needs for education in technical communication and what advice they had for program administrators, many commented on the need for programs to encourage academic minors in business and leadership as well as a technology-focused discipline, such as computer science.

The need for students to be prepared to work in business environments was a key theme in our conversations—industry leaders, in particular, talked about the importance of students having opportunities to learn about business management, strategy, and processes and to practice skills such as business analysis and business process modeling. The need for students to be prepared to work in product development environments and to understand how product development processes are connected to business processes was also a key theme. As we have demonstrated throughout this book, technical knowledge and skills are desired more than any other skill set for entry-level content-focused positions. This includes an eagerness to learn how different tools work, how to work in different kinds of application environments, and how to move between and among tools to achieve different goals. Importantly, according to many of the industry leaders with whom we spoke, this also includes AI and code literacy, familiarity with markup languages, hardware and software knowledge, and, preferably, expertise in a technical domain.

Completing one or more minors in business, computer science, and/or engineering is an ideal way for students preparing for careers as content professionals to gain these desired and valued skill sets. Students interested in working in the finance industry or science domains might consider minors in those areas as well.

- **Prioritizing project-based learning and cross-disciplinary team-based learning.** Whereas elective courses and minors in other disciplines help students develop competencies that inform their understanding of the larger systems and processes in which content workflows are situated, team-based, cross-disciplinary projects can help students develop and apply the seven competencies—systems knowledge and thinking, rhetorical proficiency, abstraction, research, social proficiency, experimentation, and technological literacy—deemed important for early career content professionals.

  Students in an advanced technical communication course focused on structured authoring and content management, for example, might be paired with students in a project-based software development course. Small teams could be assigned to create a product that meets the needs of a particular market. They would ideally engage in the full product life cycle, from research to planning to developing to testing—with students documenting every step of the process along the way, including documenting the product itself using a structured authoring approach. Students would employ Agile development methods and a Scrum framework to carry out daily activities, and they would help each other learn disciplinary approaches to software development (computer science students) and product documentation (technical communication students) along the way. Should teams decide to use GitHub to build and document the software, computer science students could help technical communication students troubleshoot as needed.

In response to our question about advice for academic administrators, several industry leaders recommended providing students more opportunities to engage in "end-to-end experiences," or projects that engage students in the full product life cycle and that require collaborative, cross-disciplinary problem-solving. An information architect and CEO of a communications consultancy talked about the value of technical communication and computer science faculty team teaching courses or joint projects that brought technical communication and computer science students together. He envisioned students engaging in problem-solving throughout a product life cycle:

> It would be good to have those projects start at the beginning of requirements. Take them through the product life cycle. What's the business problem? When the customer orders the product, what happens? What humans have to intervene? What's the output from that process? And so on. And can you diagram it? Again, we're getting at process and within process, little procedures. Some of them done by software; some

> of them done by humans.... Again, that, I think, would help them to do soup to nuts on business processes.

In addition to gaining a better understanding of how business processes connect with both product and content life cycles, and how content development syncs with Agile workflows, technical communication students who engage in team projects with computer science students would become better versed in computer science concepts and learn—likely through much trial and error—strategies for collaborating with subject matter experts and advocating for designs and content that meets end-user needs. Such end-to-end experiences can offer students rich opportunities to develop and hone the competencies content professionals need to thrive.

Professional Development

Lack of access to training programs and materials in the disciplinary areas of content is a significant challenge facing most educators in the field of technical communication. So too are a lack of resources and time to support additional knowledge and skills development. At the 2019 Association of Teachers of Technical Writing (ATTW) conference, where we presented on a panel with Tracy Bridgeford and others who had published entries in the collection *Teaching Content Management in Technical and Professional Communication* (2020a), we called this professional development/training problem the "elephant in the field."

Bridgeford (2020b) aptly described many of the challenges facing technical communication educators who may be interested in developing new content-focused knowledge and skills so to innovate curricula and pedagogy:

> Many instructors teach 4/4 loads, and are not provided with professional development time (or compensation) for curriculum work, and therefore may have limited time or means to gain the knowledge needed to rethink curriculum and pedagogy. There simply is not enough time to learn, to adapt, and to redesign curricula and pedagogies, especially when these processes also require that they learn new technologies. But change we must, no matter how comfortable or time-consuming. We must evolve, update, and rethink our curricular frameworks and pedagogical habits as technological changes affect the field.
>
> *(p. 2)*

> I believe for many of us teaching technical communication, even today, single sourcing is still a foreign concept not because instructors

> refuse to change, but because they lack opportunities for professional development.
>
> (p. 5)

We view the need for professional development in the disciplinary areas of content as a pressing issue, one that both industry and academic leaders in the content-focused facet of the field need to come together to address. Professional development opportunities focused on teaching are also needed.

Below, we offer ideas for ways in which educators can learn more about the disciplinary areas of content and begin to develop foundational knowledge and skills that support teaching and curricular development. We offer these ideas with the understanding that most educators are dealing with limited time, resources, and institutional support. An initial version of these ideas appeared in our afterword (Evia & Andersen, 2020) to Bridgeford's (2020a) collection.

- **Participate in a learning community.** Learning communities for educators that focus on the discipline of content would be immensely valuable resources, and we want to propose the idea here as something worthy of our professional organizations pursuing. We know all too well that a majority of programs offering degrees in technical communication have only a handful of faculty whose research and teaching are dedicated to technical and professional communication. And even fewer have faculty who feel well prepared to teach courses and curricular units focused on modern technical content practices. This leaves those who are interested in learning approaches to structured authoring, content modeling, content management, and other content-focused practices often feeling isolated and unsupported. Faculty in the field need access to learning communities dedicated to professional development in the content-focused facet of the field, and our professional organizations seem best positioned to create and sustain these.

  We imagine organizations such as the CPTSC or the ATTW sponsoring learning communities for educators interested both in developing their own knowledge and skills and in developing courses and curricular units focused on the knowledge and practices associated with the disciplinary areas of the content discipline. These communities would, above all, support learning and perhaps might be viewed as virtual maker spaces, where members collaboratively problem-solve as they make content things and build curricula. Regular meetings, perhaps bi-monthly, would allow for members to build knowledge and skill sets continually and to share successes and challenges as they work on projects. Members with expertise in particular practices might lead workshops, assign projects, and offer guidance and feedback over several meetings.

A learning community connected to the work of the SIGDOC Structured Authoring Committee has much promise. As we mentioned before, the committee is working to develop free resources to support educators who want to teach structured authoring at the assignment, course, or program level. The resources include both conceptual (explanatory) and procedural (tutorial) modules that are intended to be adapted for the classroom. The committee has also discussed starting a workshop series, which could also offer the learning community guidance and support.

- **Complete free online training.** Scriptorium hosts a series of free DITA courses at learningdita.com, which we mentioned earlier in this chapter. These are excellent training resources for both educators and students. The courses move from introducing DITA basics to creating DITA topic types to assembling, reusing, and publishing DITA content.

  The help documentation of different CCMSs and XML authoring tools can also be a great training resource. Educators can download trial versions of most tools and follow the documentation to learn how to create, filter, reuse, publish, and manage structured content. We particularly recommend as a training resource the documentation of the Heretto CCMS and Oxygen XML Editor.
- **Participate in industry webinars, workshops, and online communities.** The CIDM, Content Wrangler, Data Conversion Laboratory (DCL), Heretto, and Scriptorium offer free and frequent webinars focused on topics of interest in the discipline of content. The webinar series "Coffee and Content" hosted by The Content Wrangler and "Let's Talk Content Ops!" hosted by industry expert Sarah O'Keefe are particularly robust and informative. The DCL Learning Series is also a great, free resource with an instructional focus.

  Comtech Services and Single-Sourcing Solutions offer regular online workshops and courses focused on a variety of content practices, from DITA training to incorporating AI into a content strategy to advanced reuse strategies, but these are at a cost. We encourage educators to reach out to the leadership teams of Comtech and Single-Sourcing Solutions to inquire about discounts for educators for particular workshops and courses. The Write the Docs community is also an active and welcoming community to follow—members are dedicated to improving the art and science of software documentation. The community has regular meetups in multiple cities each month, and members communicate frequently through Slack channels.

  Our academic-oriented professional organizations might consider offering more preconference workshops at the annual conferences.

Organizers might invite educators or industry professionals who are experts in content practices such as structured authoring, content modeling, and taxonomy development to offer workshops (a portion of workshop fees could go toward compensation).

- **Follow podcasts dedicated to important topics in the discipline of content.** Numerous podcasts are dedicated to covering a wide range of topics situated within the discipline of content. We recommend the following podcasts:

    - *Content+AI* hosted by Larry Swanson
    - *The Content Strategy Experts* hosted by Scriptorium
    - *Content Matters* hosted by Ingeniux
    - *Content Components* hosted by Heretto
    - *The Content Strategy Podcast* hosted by Kristina Halvorson.

- **Seek out teaching-focused scholarship that describes pedagogical strategies and activities for teaching different practices in the discipline of content.** While teaching-focused scholarship focused on the discipline of content is limited, there are a handful of excellent books, chapters in edited collections, and articles that offer ideas and guidance on how to teach specific practices, such topic-based authoring, structured authoring, intelligent content creation with Lightweight DITA, and content strategy. In addition to the resources that we listed in the different stages presented in the "Building Curricular Pathways" section of this chapter, we particularly recommend the following publications:

    - "Writing about Structure in DITA" in *Teaching Content Management in Technical and Professional Communication* by Jason Swarts (2020, pp. 155–173)
    - "Rhetorically Structured Content: Developing a Collaborative Single-sourcing Curriculum" in *Technical Communication Quarterly* (17.1) by Charlotte Robidoux (2008, pp. 110–135)
    - "Teaching Topic-based Writing" in *Teaching Content Management in Technical and Professional Communication* by Yvonne Cleary (2020, pp. 72–86).

We would be remiss not to talk about the training of graduate student instructors in this section on professional development. Because graduate students in technical and professional communication MA/MS and PhD programs often land academic positions that require them to teach, among other courses, the technical writing service course and/or courses in an undergraduate technical communication major, minor, or certificate program, educators who teach pedagogy and research seminars as well as practicums should consider ways to integrate readings, discussions, and assignments focused on the discipline of content.

In pedagogy seminars focused on preparing graduate students to teach the undergraduate technical writing service course, for example, graduate students might be assigned more general readings focused on the content discipline and its drivers and more specific readings focused on the practice of structured authoring, which is most relevant to students in the technical writing service course. Any of the resources listed above would be appropriate to assign in a graduate seminar focused on pedagogy or research or in a practicum focused on developing skills in content design and content strategy. Ideally, graduate students would have an opportunity to complete, at the very least, a small-scale documentation project in which they apply structured authoring methods, standards, and tools; such a project would help them gain enough knowledge and know-how to be able to introduce future students to structured authoring approaches.

Conclusion

As we wrap up this project, we can say that working at the intersection of technical communication and the discipline of content is not as simple as following one project assignment or attending a conference workshop. Those are undeniably useful initial stages, but this intersection is complex, dynamic, and evolving, and it involves participants from academia, industry, nonprofits, and government offices that work with technical content. In Chapter 1, we wrote that exploring curricular pathways that prepare students as content professionals does not mean that technical communication programs in academia need to re-invent themselves or abandon their existing strengths in other facets of the field. However, we have pointed out throughout this book that, for better or worse, whole sectors of the workplace think of the type of technical content that we have described whenever the term "technical communication" comes to mind—and we should not ignore that. Many managers and leaders across those workplace areas are actively looking for a new generation of technical content professionals.

To achieve a balance of demand and supply for that new generation of technical content professionals, those of us working in industry and in academia have work to do. We urge our colleagues in industry to consider the following goals that we outlined in Chapter 4:

- *Reach some consensus to define what an entry-level content-focused position looks like in technical content.* Also explain what experience and qualifications can realistically be expected from college graduates with degrees in technical communication and related fields.
- *Develop onboarding training programs.* These can establish when the expected academic preparation of a candidate ends and where structured on-the-job training begins for specific positions or job titles.

- *Articulate career paths for technical content professionals.* We need to know what entry-level positions entail and what is reserved or expected of mid- or advanced-career professionals in order to create or revise curricula at the undergraduate, post-baccalaureate, graduate, and certificate levels.
- *Establish and maintain partnerships with universities based on the skill shapes needed in different regions, sectors, and professions.* These partnerships should be grounded on realistic expectations (university degrees are not just job preparation ramps) and require active involvement. These collaborations can extend to participation on advisory boards and in internship and career fairs, presentations in college classrooms, and recruitment of students to attend industry conferences, webinars, and workshops. For faculty, these collaborations can inject ideas into curriculum design and professional development activities.
- *Expand representation and recruitment efforts in career and internship fairs.* Some managers who have approached us at conferences complaining about the scarcity of technical communication graduates work for companies that actively recruit in our universities for positions related to business and engineering. Consolidated recruitment efforts should include areas related to technical communication and the discipline of content.

Our colleagues in academia interested in the intersection of technical communication and the discipline of content can take action in some of the following ways:

- *Embrace a lifelong learning approach to technical communication education.* A degree program in a conventional university environment can only go so far in preparing students for careers beyond entry-level positions. Consider the importance of experience and recalibrate the scope of academic training. Explore "blended programs" at the undergraduate, graduate, and certificate levels that build on a student's domain knowledge, training, and experience.
- *Recognize the work of "lone ranger" faculty who in many cases are the only technical content instructors in a program.* Students and colleagues often look up to these instructors as opinion leaders and mentors, and that can be an added tax on expected workloads. If the regional or industry demands of a given program demand more technical content professionals, these lone ranger faculty members will need peers, which can also come from industry: professors of practice can contribute to curricular efforts in unique ways.
- *Reconsider the use of the term "technical communication" in university programs based in the United States.* At the program level, and even

field level, the teaching and research of what we call technical communication in this country covers a large and diverse (in some cases disparate) landscape. This narrow term for such a large collection of activities and knowledges can be confusing for students, academics, and practitioners beyond our field. "Technical communication" in the United States comprises many facets or components (refer to Melonçon & Schreiber, 2022) that do not necessarily reflect the usage of this term outside of US-based academic environments.

- *Establish and maintain partnerships with industry leaders.* Particularly in regional communities, this should address the specific skill shapes of the profession in a given context. Connect these outreach efforts to high school populations to create a united industry + academia front that introduces technical content to students before they go to college or university.
- *Connect facets of technical communication around the discipline of content.* In Chapter 1, we affirmed that the discipline of content can benefit from work that investigates content practices from a social justice perspective with the goal of informing practice and enacting change. Academic programs need course modules focused on structured authoring, minimalism, or taxonomy development connected to the facet of social justice. We need to teach students about ways in which the content life cycle must generate outputs that are inclusive, equitable, and serve the diverse needs of those who use them. Academia cannot just focus on cultivating good writers and designers. We need students to understand the content landscape and how these systems work in order to know, ask, and answer appropriate ethical and philosophical questions.
- *Build and sustain community efforts around the discipline of content.* Find commonalities related to pedagogical practices, challenges, and curriculum design. The ACM SIGDOC Structured Authoring Committee that we mentioned earlier in this chapter could be a starting point for many colleagues in academia.

Lastly, the recommendations included in this chapter (and the whole book) aim to align academic content with status quo practices in some sectors of industry. The dynamic discipline of content of course won't stay like this forever, and this is where academics and practitioners must collaborate in research projects that can expand our knowledge in some (and more!) of the following areas, which we present in no particular order:

- *Expand technical content to multimedia platforms.* The planned 2.0 version of the DITA standard includes native element components for audio and video that must be researched as complementing (or replacing for some audiences and contexts) mostly written technical content.

- *Explore additional intersections with the discipline of content.* We have mentioned that sometimes students come to our "content" courses expecting social media products. We should focus on research that bridges that gap and reaches new audiences for technical content. Professionals (creators, strategists, marketers) in other areas of the discipline of content face some of our challenges, as their posts and reels have to compete for users' attention with pranks and recordings that belong in the world of America's Funniest Videos. Their work, thus, is also seen as ephemeral and lightweight, even when it is produced by serious teams. Maybe a future direction for technical content is intersecting with social media content.
- *Focus on accessibility of all technical content.* The US Department of Justice made in 2024 a final rule updating its regulations for Title II of the Americans with Disabilities Act. The final rule has specific requirements about how to ensure that digital content is accessible to people with disabilities. This ruling includes technical content.
- *Write the next chapter that integrates AI into technical content workflows.* As we write this book, the technical communication community has just started to understand and adapt generative AI tools in the content life cycle. We need research collaborations that incorporate AI tools in effective, ethical, and fair ways to automate activities related to technical content.
- *Embrace mobile-first for global inclusion.* Many workflows and tools that process technical content in DITA and even docs-as-code environments prioritize PDF or desktop-aimed websites as their main deliverables. As most of the world, particularly in the Global South, accesses information primarily on mobile devices, we need research that places mobile as the first mode of consumption of multimedia technical content.
- *Inject data into the content life cycle.* At all quadrants of the life cycle, decisions can and should be supported by data to maximize resources and make precise personalization for specific audiences and contexts.
- *Integrate semantics and structured content into AI workflows.* Recent conversations have highlighted the importance of semantic content for improved training and communication with large language models. Abel (2025) compares structured content to semantics with the following analogy:

> structured content is a beautifully organized cookbook written in XML. Semantic content is the same cookbook, but with notes in the metadata that say, 'This recipe pairs perfectly with that one,' or 'Use this ingredient as a substitute in case of dietary restrictions.' It's all about adding meaning to the structure, making the content more intelligent and thus more informative.

Semantic content involves working with knowledge graphs, taxonomies, and ontologies beyond what static structure from, say, a DITA template can achieve.

- *Examine digital transformation diffusion processes.* Research that draws on rhetorical, socially constructed models of technology diffusion and adoption can help unpack cultural dynamics characterizing change initiatives and impacting the success of planned organizational and technological change. Results can help initiative leaders plan for more critical approaches to digital content transformation.
- *Explore collaborative practices in Agile environments.* Research that examines how content professionals collaborate with software development teams, product managers, and other stakeholders are needed to identify practices that work well and not so well. Results can lead to new or refined practices and to innovations in how we teach collaboration.
- *Expand audience analysis models to include audience touchpoints in the content life cycle.* Audiences who shape the design and use of content in the different stages of the life cycle include, but are not limited to, customers and users, managers and C-level executives, and cross-disciplinary team members, from content strategists to software developers to marketing specialists to UX designers.
- *Develop new or refine existing approaches for centering users.* This must happen in content models, taxonomies, and designs with the aim of providing users seamless, consistent, equitable, accessible, and memorable content experiences. Examine ways in which social justice frameworks can inform strategies and practices for centering users, particularly those who are marginalized and underrepresented.

We advocate for new research focused on the discipline of content that embraces diverse perspectives and methods, with the goal of informing future practices; questions around disciplinarity; and pedagogical and curricular approaches. Research ideas such as those presented here might be examined through perspectives and methods situated within technical communication (e.g., social justice, design thinking, discursive infrastructures) and/or within other fields, such as marketing, psychology, and software development.

Importantly, dissemination of research results to both academic and practitioner audiences will be key. Rebekka and JoAnn Hackos (Andersen & Hackos, 2021) found in their study of practicing technical communicators' experiences with and perspective on academic publishing that "academic research in the field—in both its design and reporting—has potential to reach and impact much wider audiences" (p. 46). We see new research collaborations and studies focused on the people, processes, and technologies comprising the discipline of content as a real opportunity to advance theory, contribute new knowledge, improve practices, and define and continue to refine educational pathways in the discipline.

Technical communication and the discipline of content are inextricably linked, requiring ongoing dialogue between academia and industry. As both sides of the field evolve, the challenge is not just to keep pace but to actively shape the future of technical content work—ensuring it remains inclusive, dynamic, and indispensable.

Notes

1 https://github.com/VT-Evia/content-discipline.
2 https://www.acm-sigdoc-structured.org/.
3 https://www.markdowntutorial.com/.
4 https://learningdita.com/.
5 https://www.dita-ot.org/.
6 https://pandoc.org/.

References

Abel, S. (2025, January 21). *Why technical writers must evolve from structured to semantic content*. Retrieved from https://www.thecontentwrangler.com/p/why-technical-writers-must-evolve.

Andersen, R. (2011). Component content management: Shaping the discourse through innovation diffusion research and reciprocity. *Technical Communication Quarterly*, *20*(4), 384–411.

Andersen, R. (2014). Rhetorical work in the age of content management: Implications for the field of technical communication. *Journal of Business and Technical Communication*, *28*(2), 115–157.

Andersen, R., & Batova, T. (2015). The current state of component content management: An integrative literature review. *IEEE Transactions on Professional Communication*, *58*(3), 247–270.

Andersen, R., & Hackos, J. (2021). Practicing technical communicators' experiences with and perspectives on academic publishing. *Technical Communication*, *68*(3), 29–55.

Bailie, R., & Evia, C. (2024). Operationalizing content creation. In C. Evia (Ed.), *Content operations from start to scale: Perspectives from industry experts* (pp. 43–74). Blacksburg: Virginia Tech Publishing.

Bailie, R., & Urbina, N. (2013). *Content strategy: Connecting the dots between business, brand, and benefits*. Laguna Hills, CA: XML Press.

Bridgeford, T. (Ed.). (2020a). *Teaching content management in technical and professional communication*. New York: Routledge.

Bridgeford, T. (2020b). Content management: A pedagogical exigency. In T. Bridgeford (Ed.), *Teaching content management in technical and professional communication* (pp. 1–15). New York: Routledge.

Bridgeford, T., & St.Amant, K. (Eds.). (2015). *Academy-industry relationships and partnerships: Perspectives for technical communicators*. Amityville, NY: Baywood Publishing Company.
Carey, M., Longo, D., & Lanyi, M. (2014). *Developing quality technical information: A handbook for writers and editors* (3rd ed.). Upper Saddle River, NJ: IBM Press.
Casey, M. (2023). *The content strategy toolkit: Methods, guidelines, and templates for getting content right* (2nd ed.). Hoboken, NJ: New Riders.
Cleary, Y. (2020). Teaching topic-based writing. In T. Bridgeford (Ed.), *Teaching content management in technical and professional communication* (pp. 72–86). New York: Routledge.
Department of English, Carnegie Mellon (2025). Technical Writing Program. Retrieved from https://www.cmu.edu/dietrich/english/academic-programs/technical-writing/index.html.
Duin, A., & Tham, J. (2018). Cultivating code literacy: Course redesign through advisory board engagement. *Communication Design Quarterly Review*, *6*(3), 44–58.
Duin, A., Tham, J., & Pedersen, I. (2020). The rhetoric, science, and technology of 21st century collaboration. In M. Klein (Ed.), *Effective teaching of technical communication: Theory, practice, and application* (pp. 169–192). Fort Collins, CO: The WAC Clearinghouse.
Dush, L. (2015). When writing becomes content. *College Composition and Communication*, *67*(2), 173–96.
Eichhorn, K. (2022). *Content*. Cambridge: The MIT Press.
Evia, C. (2018). *Creating intelligent content with Lightweight DITA*. New York: Routledge.
Evia, C. (2024). Introduction. In C. Evia (Ed.), *Content operations from start to scale: Perspectives from industry experts* (pp. 1–12). Blacksburg: Virginia Tech Publishing.
Evia, C., & Andersen, R. (2020). Beyond management: Understanding the many forces that shape content today. In T. Bridgeford (Ed.), *Teaching content management in technical and professional communication* (pp. 213–231). New York: Routledge.
Evia, C., Sharp, M., & Pérez-Quiñones, A. (2015). Teaching structured authoring and DITA through rhetorical and computational thinking. *IEEE Transactions on Professional Communication*, *58*(3), 328–343.
Getto, G., Labriola, J., & Ruszkiewicz, S. (2023). *Content strategy: A how-to guide*. New York: Routledge.
Hackos, J. (2011). *Introduction to DITA: A user guide to the Darwin Information Typing Architecture including DITA 1.2* (2nd ed.). Comtech Services, Inc.
Halvorson, K., & Rach, M. (2012). *Content strategy for the web* (2nd ed.). Berkeley, CA: New Riders.
Hart-Davidson, B. (2012, October 17). *Code? not so much*. Digital Rhetoric Collaborative. Retrieved from https://www.digitalrhetoriccollaborative.org/2012/10/17/code-not-so-much/.
Jones, C. (2019). *The content advantage (Clout 2.0)*. San Francisco, CA: New Riders.

Jorgensen, E. (2023). *Strategic content design: Tools and research techniques for better UX*. New York: Rosenfeld Media.

Katz, S. (2015). Creating bridges with internships. In T. Bridgeford & K. St. Amant (Eds.), *Academy-industry relationships and partnerships: Perspectives for technical communicators*, (pp. 77–96). Amityville, NY: Baywood Publishing Company.

Kramer-Simpson, E. (2018). Feedback from internship mentors in technical communication internships. *Journal of Technical Writing and Communication, 48*(3), 359–378.

Lauren, B. (2018). *Communicating project management: A participatory rhetoric for development teams*. New York: Routledge.

Lawrence, H., Fernandez, L., Garrison-Joyner, V., Bonem, M., Eyman, D., Lussos, R., & Schafer, L. (2023). Sustainable industry-university partnerships: Accountability as a model for program development. *Programmatic Perspectives, 14*(1), 41–71.

MacIntyre, J. (2024). Personalization and content operations. In C. Evia (Ed.), *Content operations from start to scale: Perspectives from industry experts* (pp. 91–108). Blacksburg: Virginia Tech Publishing.

Melonçon, L., & Schreiber, J. (2022). Introduction: Promoting a sustainable collective identity for technical and professional communication. In J. Schreiber & L. Melonçon (Eds.), *Assembling critical components: A framework for sustaining technical and professional communication* (pp. 3–16). WAC Clearinghouse: University Press of Colorado.

Mendenhall, A. (2013). The historical problem of vertical coherence: Writing, research, and legitimacy in early 20th century rhetoric and composition. *Composition Studies, 41*(1), 84–100.

Pringle, A., & O'Keefe, S. (2009). *Technical writing 101: A real-world guide to planning and writing technical content*. Durham, NC: Scriptorium Press.

Quetzalli, A. (2023). *Docs-as-ecosystem: The community approach to engineering documentation*. New York: Apress.

Quigley, S. J., Lui, E., Whelpley, S., & Flot, J. (2022). Writing infrastructures: GitHub in the technical and professional communications classroom. *Reflections, 22*(1), 242–274.

Robidoux, C. (2008). Rhetorically structured content: Developing a collaborative single-sourcing curriculum. *Technical Communication Quarterly, 17*(1), 110–135.

Rockley, A., Cooper, C., & Abel, S. (2015). *Intelligent Content: A Primer*. Laguna Hills, CA: XML Press.

Rude, C. (2015). Considering partnerships and relationships in the field. In T. Bridgeford & K. St. Amant (Eds.), *Academy-industry relationships and partnerships: Perspectives for technical communicators* (pp. v–ix). Amityville, NY: Baywood Publishing Company.

Rushkoff, D. (2011). *Program or be programmed: Ten commands for a digital age*. New York: Soft Skull Press.

Sides, C. (2021). Internships in a pandemic. *Programmatic Perspectives, 12*(1), 1–21.

Smith, H. (2015). Collaborating with industry using mentoring programs and internships. In T. Bridgeford & K. St. Amant (Eds.), *Academy-industry relationships and partnerships: Perspectives for technical communicators* (pp. 97–116). Amityville, NY: Baywood Publishing Company.

Swarts, J. (2020). Writing about structure in DITA. In T. Bridgeford (Ed.), *Teaching content management in technical and professional communication* (pp. 155–173). New York: Routledge.

Technical Communication Advisory Board. (2024). College of Liberal Arts Writing Studies, University of Minnesota. Retrieved December 17, 2024, from https://cla.umn.edu/writing-studies/alumni-friends/technical-communication-advisory-board.

Weise, M. R. (2021). *Long life learning: preparing for jobs that don't even exist yet*. Hoboken, NJ: John Wiley & Sons, Inc.

White, L.W. (2016). Single sourcing. In R. Gallon (Ed.) *The language of technical communication* (pp. 56–57). Laguna Hills, CA: XML Press.

Why your company is struggling to scale up generative AI. (2024, November 4). *The Economist*. Retrieved February 4, 2025, from https://www.economist.com/business/2024/11/04/why-your-company-is-struggling-to-scale-up-generative-ai.

Willerton, R. (2015). *Plain language and ethical action: A dialogic approach to technical content in the 21st century*. New York, NY: Routledge.

INDEX

Note: **Bold** page numbers refer to tables and *italic* page numbers refer to figures.

For Product Safety Concerns and Information please contact our EU
representative GPSR@taylorandfrancis.com
Taylor & Francis Verlag GmbH, Kaufingerstraße 24, 80331 München, Germany

www.ingramcontent.com/pod-product-compliance
Ingram Content Group UK Ltd.
Pitfield, Milton Keynes, MK11 3LW, UK
UKHW021723240926
475702UK00030BA/2179

* 9 7 8 1 0 3 2 5 8 8 4 6 9 *